Teaching the Entrepreneurial Mindset to Engineers

Lisa Bosman · Stephanie Fernhaber

Teaching the Entrepreneurial Mindset to Engineers

Springer

Lisa Bosman
Milwaukee, WI
USA

Stephanie Fernhaber
Indianapolis, IN
USA

ISBN 978-3-319-61411-3 ISBN 978-3-319-61412-0 (eBook)
DOI 10.1007/978-3-319-61412-0

Library of Congress Control Number: 2017946038

Printed on acid-free paper

This Springer imprint is published by Springer Nature
The registered company is Springer International Publishing AG
The registered company address is: Gewerbestrasse 11, 6330 Cham, Switzerland

Acknowledgements

We would like to thank the many people who assisted with this book, including everyone who provided support, offered feedback, assisted in the editing, and contributed exercises. We are especially appreciative to the following individuals: Dr. Casey Allen, Dr. Elise Barrella, Dr. Samuel Bechara, Dr. Cristi Bell-Huff, Dr. Anthony Bowman, Dr. Bethany Brinkman, Dr. Dennis Brylow, Dr. Keith Buffinton, Dr. Ken Christensen, Dr. Andrew Gerhart, Dr. Jay Goldberg, Dr. Jing Guo, Dr. David Jamison, Dr. Susana Lai-Yuen, Dr. Brooke Mayer, Dr. Patrick McNamara, Dr. Douglas Melton, Marta Moldvai, Heidi Morano, Dr. Hyunjae Park, Dr. Kurt Paterson, Dr. Kris Ropella, Dr. John Santiago, Dr. Robert Scheidt, Dr. Rudy Schlaf, Dr. Taly Schmidt, Dr. Michael Swartwout, Dr. Massoud S. Tavakoli, Dr. Nathan Weise, Karen Wilken, Dr. Julia M. Williams, and Dr. Jose L. Zayas-Castro.

Contents

About the Authors

Lisa Bosman, Ph.D. develops and facilitates faculty professional development experiences to incorporate the entrepreneurial mindset into the engineering classroom within the Opus College of Engineering at Marquette University. In addition, Lisa teaches a variety of engineering courses at the undergraduate and graduate levels, provides consulting related to the entrepreneurial mindset and best teaching practices, has received numerous grant awards to improve STEM curriculum, and is active within the Milwaukee start-up community. Prior to joining higher education, she worked with Harley-Davidson Motor Company, John Deere, and Oshkosh Corporation (formerly, Oshkosh Truck). Lisa received her Ph.D. in Industrial Engineering from the University of Wisconsin–Milwaukee.

Stephanie Fernhaber, Ph.D. is an Associate Professor of Entrepreneurship within the Lacy School of Business at Butler University. She teaches a variety of entrepreneurial courses at the undergraduate and graduate levels including The Entrepreneurial Mindset, First-Year Business Experience, Real Business Experience, and Social Entrepreneurship. In addition to teaching, Stephanie conducts research in the areas of international entrepreneurship, networks and new venture strategy. Her background includes a combination of consulting, corporate finance, and entrepreneurial initiatives. Stephanie received her Ph.D. in Entrepreneurship from Indiana University.

Chapter 1
Introduction

Entrepreneurs are everywhere.—Eric Reis, The Lean Startup.

Entrepreneurship is all around us. We watch entrepreneurs pitch their ideas on Shark Tank. We read stories in our local newspapers and hear how accelerator programs, sponsored by organizations such as the Midwest Energy Research Consortium in Milwaukee, are helping launch new start-ups. We likely have friends or family who are entrepreneurs, and if not, we often read about individuals such as Steve Jobs, Sara Blakely, Madam C.J. Walker, or Elon Musk. We hear about how large, existing companies, such as Google or 3M, go to great efforts to create cultures of entrepreneurship for their employees. Within major business publications such as the *Wall Street Journal*, entrepreneurship and innovation is becoming part of the everyday conversation. We read about how established social enterprises such as Goodwill Indy are innovating and expanding with new educational and healthcare solutions for addressing poverty. Even government entities appear to be relying on entrepreneurship, incorporating innovations such as Street Bump to detect potholes or creating its own co-working spaces. This reliance on entrepreneurship and innovation all around us coincides with an increasing acceleration of technological change, often resulting in the disruption of industries and the emergence of new ones. This is the case of eBay, Uber, and 3D printing. Indeed, we are in the midst of an entrepreneurial revolution where uncertainty and change have become the norm.

Engineering students are embracing these changes as well. The Kaufman Foundation (2011) reports that over 50% of 18–24 year olds want to start their own business. Currently, entrepreneurship is a major found within the business school, leading students to choose between engineering and entrepreneurship as opposed to exploring them together. Alternatively, engineering students that are interested in entrepreneurship can visit a university entrepreneurship center or take extra coursework through the business school, if it fits into their schedule. Yet, this solution seems to suggest that developing an entrepreneurial mindset is optional and can be done in your spare time *if* it is convenient to do so. We believe that the entrepreneurial mindset is a necessity for engineers, and agree with the Royal Academy of Engineering (2012) that entrepreneurship must "be presented as an

L. Bosman and S. Fernhaber, *Teaching the Entrepreneurial Mindset to Engineers*, DOI 10.1007/978-3-319-61412-0_1

integral part of the curriculum." Entrepreneurship is the backbone of the global economy, and engineers play a critical role by addressing pressing problems and creating new and innovative solutions. This is not a role that is played in isolation, but rather requires collaboration and leadership. It is a role that involves technical skills, but also requires curiosity, an ability to connect pieces of knowledge to discover solutions, and a focus on value creation—which are all characteristic of an entrepreneurial mindset.

A survey of engineering students by Duval-Couetil et al. (2012) found that two-thirds of the students thought entrepreneurship education could broaden their career prospects and choices. Surprisingly, only one-half of the engineering faculty surveyed considered access to entrepreneurship education as valuable. This hints at part of the reason why integrating entrepreneurship into the engineering curriculum has been relatively slow, with a few notable exceptions, and why some initiatives have been student demand driven. Indeed, this has been the approach of the University Innovations Fellows program that empowers students to go back to their institutions and become "innovation ambassadors." Nonetheless, there is an exciting movement brewing around the integration of entrepreneurship education into engineering as evidenced by the mission and work of the Epicenter at Stanford, the Kern Entrepreneurial Engineering Network (KEEN), and a recent special issue in the *Advances in Engineering Education* journal specifically focusing on entrepreneurship and innovation. Yet, while the entrepreneurial mindset is increasingly being accepted as important within the engineering arena (Byers et al. 2013), the movement still has a long ways to go in order to reach all engineering students.

It is rare for engineering faculty to be trained in entrepreneurship. Therefore, it is critical to equip engineering educators with tools and strategies that allow them to teach the entrepreneurial mindset. How might the entrepreneurial mindset be integrated into the engineering fundamental courses? What are the best practices? What resources and support exist? How do I justify my approach to my students, peers and administration? What are the next steps? These are just a few of the questions we address in this book. Our goal is quite simple. We want to advance entrepreneurship education for *all* engineering students, and we need to motivate and equip you, as the engineering educator, to do so.

Our goal is simple. We want to advance entrepreneurship education for all engineering students, and we need to motivate and equip you, as the engineering educator, to do so.

This book is broken down into four parts. In the first section, we dig deeper to explore just what the entrepreneurial mindset is and why it is important. If you are not yet convinced or need help getting others convinced, this section will provide valuable insight. Within the second section, we address the question of "how" to get started. We focus exclusively on ways in which the entrepreneurial mindset can be integrated into existing coursework because this is where you, the engineering educator, can most immediately implement change. The third section serves as a resource guide, in providing guidance as to the growing multitude of conferences, journals, networks, and online resources out there for you to rely on. Some of these resources are more general in the entrepreneurial sense, while others are custom to engineering. We also help "make the case" for integrating the entrepreneurial mindset at your university. There are many stakeholders involved and getting buy-in is key. The fourth and final section provides some solid examples to get you started.

We invite you to read this book, and better yet apply its content, examples, and resources to your engineering classroom. With any luck, you might even further develop your own entrepreneurial mindset in the process!

References

Byers, T., T. Seelig, S. Sheppard, and P. Weilerstein. 2013. Entrepreneurship: Its role in engineering education. *The Bridge on Undergraduate Engineering Education* 43 (2): 35–40.

Duval-Couetil, N., T. Reed-Rhoads, and S. Haghighi. 2012. Engineering students and entrepreneurship education: Involvement, attitudes and outcomes. *International Journal of Entrepreneurship Education.* 28 (2): 425–435.

Kaufman Foundation. 2011. Young invincibles policy brief. Retrieved from: http://www.kauffman.org/~/media/kauffman_org/research%20reports%20and%20covers/2011/11/millennials_study.pdf.

Royal Academy of Engineering. 2012. Educating engineers to drive the innovation economy. London.

Part I
What is the Entrepreneurial Mindset and Why is it so Important for Engineers?

Chapter 2
Defining the Entrepreneurial Mindset

Author Reflection

At the beginning of the semester, I often ask my students to share their thoughts on what an entrepreneur is. Student responses commonly fall into one of three categories. The first has to do with someone being a "business owner." After some discussion, this is usually further refined to "small business owner," with the caveat that the owner must be actively involved in the day-to-day operations of the business. Indeed, most of us know someone who owns his or her own business. The second category of responses typically defines an entrepreneur as someone who is "starting up their own business." Thus, the focus shifts from ownership to the start-up process. The last category of responses commonly associates entrepreneurs with "new, high-tech firms" that grow quickly, obtain venture capital funding and redefine industries. Examples might include Uber or Facebook. While these answers may be, in some cases, true, they do not get at the essence of what being entrepreneurial really means.

The purpose of this chapter is to provide an introduction and definition of the entrepreneurial mindset, taking into consideration literature and research from the discipline of entrepreneurship.

2.1 What is a Mindset?

What is the entrepreneurial mindset? To arrive at a definition, we must first dig into the meaning of the words "entrepreneurial" and "mindset" separately. We start by examining the definition of mindset, which serves as our foundation, and then move on to what it means to be entrepreneurial.

A mindset is defined by Merriam-Webster as a "mental attitude or inclination." As further elaborated upon by Thum (2012):

L. Bosman and S. Fernhaber, *Teaching the Entrepreneurial Mindset to Engineers*, DOI 10.1007/978-3-319-61412-0_2

> Your mindset is the sum of your knowledge, including beliefs and thoughts about the world and yourself in it. It is your filter for information you get in and put out. So it determines how you receive and react to information.Mindset: Mental attitude or inclination

In her book entitled Mindset, Carol Dweck (2006) acknowledges that a mindset can be fixed or growth oriented. While a fixed mindset assumes your talents and abilities are set, the growth mindset believes your talents and abilities can be developed. Yet, Dweck is quick to point out that a mindset can change. Much of this occurs through developing a greater awareness of your current mindset, and taking steps to purposely start thinking and reacting in new ways.

Mindset: Mental attitude or inclination

From an academic standpoint, the mindset concept comes out of the cognitive psychology and organization theory fields. Gupta and Govindrarajam (2002: 116–117) summarize the findings from this literature as they pertain to individuals:

1. **As human beings, we are limited in our ability to absorb and process information**. Thus, we are constantly challenged by the complexity, ambiguity, and dynamism of the information environment around us.
2. **We address this challenge through a process of filtration**. We are selective in what we absorb and biased in how we interpret it. The term mindset refers to these cognitive filters.
3. **Our mindsets are a product of our histories and evolve through an interactive process**. Our current mindset guides the collection and interpretation of new information. To the extent that this new information is consistent with the current mindset, it reinforces that mindset. From time to time, however, new information appears that is truly novel and inconsistent with the existing mindset. When this happens, we either reject the new information or change our mindset. The likelihood that our mindsets will undergo a change depends largely on how explicitly self-conscious we are of our current mindsets: the more the hidden and subconscious our cognitive filters, the greater the likelihood of rigidity.

Thus, while our mindsets can be shaped by an intentional awareness, they are also largely driven by our experiences and what we are (intentionally or unintentionally) exposed to.

…while our mindsets can be shaped by an intentional awareness, they are also largely driven by our experiences

What does this mean for engineering educators? For starters, we have a direct way of influencing the mindset of the students through the experiences we provide them with in our classes. Yet, as mindsets evolve through an interactive process, it is not simply a "one and done" effort. Rather, it must be reinforced and practiced. **In this sense, a mindset is a habit that requires practice**.

2.2 What Does it Mean to be Entrepreneurial?

The word "entrepreneur" originates from the French word *entrependre*, which means "undertaker" as in the sense of someone undertaking a major project. The French economist most well-known for coining this term around 1800 is Jean Baptiste Say, who emphasized "the entrepreneur shifts resources out of an area of lower and into an area of high productivity and greater yield" (Drucker 1985a: 23). Thus, the creation of value is of the very essence of entrepreneurship. Yet, the definition has continued to evolve from here, as illustrated in Table 2.1.

Joseph Schumpter is perhaps best known for coining the term "creative destruction" in 1942, which emphasizes the "new" element of entrepreneurship in that new innovations replace existing products, processes, and services. According to Schumpeter (1942), "the function of entrepreneurs is to reform or revolutionize the pattern of production … by exploiting an invention or, more generally, an untried technological possibility for producing a new commodity or producing an old one in a new way, by opening up a new source of supply of materials or a new outlet for products, by reorganizing an industry and so on." This hereby suggests that entrepreneurship can take many different forms or combinations. For example, Henry Ford installed the first assembly line for automobiles, creating a more efficient and cost-effective process. The discovery of digital cameras disrupted the photography field. With the help of software engineers, eBay brought the auction process online.

Table 2.1 Definitions of Entrepreneur(ship)

Definition	Emphasis	Origin
The entrepreneur shifts economic resources out of an area of lower and into an area of higher productivity and greater yield	Creates value	Jean Baptiste Say (1800s)
The function of entrepreneurs is to reform or revolutionize the pattern of production … by exploiting an invention or, more generally, an untried technological possibility for producing a new commodity or producing an old one in a new way, by opening up a new source of supply of materials or a new outlet for products, by reorganizing an industry and so on	Change agent/creative destruction	Joseph Schumpeter (1942)
The pursuit of opportunity without regard to resources currently controlled	Resourcefulness and uncertainty	Howard Stevenson (1983)
The entrepreneur always searches for change, responds to it, and exploits it as an opportunity	Opportunity	Drucker (1985a, b)
The discovery, evaluation, and exploitation of opportunities	Entrepreneurship as a process	Shane and Venkataraman (2000)
Entrepreneurial thinking differs from managerial or strategic thinking	Effectuation	Saras Sarasvathy (2005)

Stevenson (1983: 3) builds on the above definitions with a focus on resourcefulness, and defined entrepreneurship as "the pursuit of opportunity without regard to resources currently controlled." Entrepreneurs typically start with a constrained set of resources and need to either operate with less or find a way to mobilize external resources. This is where risk comes into play. As a new idea is not yet proven, there is an element of risk when allocating resources. This risk and uncertainty might be related to assumptions about customer desirability and demand, technological feasibility in being able to make the product or service, and/or business viability as it relates to execution, finances, etc. This is where risk management comes in.

Peter Drucker was influential in pointing out that not every small business was entrepreneurial. He also contends that the goal of entrepreneurship is not always profit—and exemplifies this notion within the public sector. Specifically, Drucker (1985b: 5–6) notes "Today, much confusion exists about the proper definition of entrepreneurship. Some observers use the term to refer to all small businesses; others, to all new businesses. In practice, however, a great many well-established businesses engage in highly successful entrepreneurship. The term, then, refers not to an enterprise's size or age but to a certain kind of activity. At the heart of that activity is innovation: the effort to create purposeful, focused change in an enterprise's economic or social potential." Indeed, this helps lay the foundation for the realization that entrepreneurship is not context specific. Drucker further emphasizes, "this defines entrepreneur and entrepreneurship—the entrepreneur always searches for change, responds to it, and exploits it as an opportunity."

Entrepreneurship: The discovery, evaluation, and exploitation of opportunities

In the year 2000, as an effort to help substantiate the academic field of entrepreneurship, Shane and Venkataraman posited that entrepreneurship seeks to understand the sources of *opportunities*; the *processes* of discovery, evaluation, and exploitation of opportunities; and the set of *individuals* who discover, evaluate, and exploit them. The "discovery, evaluation and exploitation of opportunities" is the most widely accepted definition of entrepreneurship yet today, and the definition that we adopt in this book. The authors again emphasize that entrepreneurship can include, but does not require, the creation of a new organization. Entrepreneurial opportunities are defined as new means–ends relationships. Thus, they must create value. In addition, given that these relationships are newly discovered, there is a level of uncertainty surrounding the relationships early on.

In order to better understand the way entrepreneurs think, Saras Sarasvathy interviewed 42 expert entrepreneurs as part of her dissertation. Her research concluded that entrepreneurs are distinguished not by their common traits, but rather by their common logic, or thinking process, used to solve entrepreneurial problems under conditions of uncertainty. In this way, entrepreneurial thinking is argued to differ from managerial or strategic thinking. Thus, effectuation theory was introduced. As noted by Sarasvathy (2005: 9), "Entrepreneurs are entrepreneurial, as differentiated from managerial or strategic, because they think effectually; they believe in a yet-to-be-made future that can substantially be shaped by human action; and they realize that to the extent that this human action can control the future, they need not expend energies trying to predict it."

In a recent study by Zappe and colleagues (2013), twenty-six engineering faculty that taught entrepreneurship were asked to rank the characteristics that they felt defined an entrepreneur. The top response was "acts on opportunities," which seems to suggest an alignment with the definitions provided above. Another survey of 144 engineering faculty and administrators had strong agreement on the top six choices that make up an entrepreneurial engineer (Byers et al. 2013). At the individual level, these included creativity (71%) and risk tolerance (49%). At the market level, these included opportunity recognition (64%), value creation (57%), and market awareness (47%). Within technology, the list included product development (57%). Again, these are all building on different aspects of the above definitions.

In examining the evolution of the definition of entrepreneur, it is important to recognize five key insights as they pertain to engineering education.

Insight #1 "Are you an entrepreneur?" is the wrong question. Rather, it should be "how entrepreneurial are you?"

While we have an inherent tendency to dichotomize and classify individuals as either being an entrepreneur or not—it is not necessarily this simple. As entrepreneurship is the discovery, evaluation, and exploitation of opportunities, individuals can vary as to the extent to which they exhibit these behaviors. Thus, the question shifts from "are you an entrepreneur?" to "how entrepreneurial are you?" The goal as engineering educators is to simply push our students further up the continuum, and to a place where they are more apt to discover, evaluate, and exploit opportunities on a regular basis.

Insight #2: Being entrepreneurial is relevant to more than just start-ups.

While we, again, have an inclination to associate entrepreneurship with business ownership or the high-growth, tech start-up, any reference to context is noticeably absent from the definitions. As the acceleration of change increases, the need for the discovery, evaluation, and exploitation of opportunities also intensities—regardless of what type of organization it is. While entrepreneurship has become relevant to nonprofits, government, social enterprise, and communities, it has also become significant to existing, large corporations that are working harder than ever to embrace the entrepreneurial values—as it is critical to maintaining competitiveness. This is a particularly relevant point for engineer educators, as it helps justify why entrepreneurship is applicable for all engineering students—and not just the ones that are interested in starting up their own business.

Insight #3: There are many types (and sizes) of entrepreneurial opportunities.

If entrepreneurship is defined as the discovery, evaluation, and exploitation of opportunities, there is often discussion over what constitutes an opportunity. As noted by Schumpeter, innovations can involve any new combination of knowledge, resources, etc. He provides five general types of innovations, including new goods, new methods of production, new markets, new sources of inputs, and new types of organizing. We can also look at the distinction between radical versus incremental innovations. For entrepreneurial educators, the main takeaway is that entrepreneurial opportunities can take many different shapes and sizes. It can range from developing a completely new product, for example, to refining an existing process to reach a new market.

***Insight #4: Being entrepreneurial requires the creation of value*.**

From the earliest definition of entrepreneurship provided by Jean Baptiste Say in 1800, it was clear that being entrepreneurial involves the creation of value. This value can come in many different forms, whether it is a civil engineer improving the traffic flow at an intersection or a biomedical engineer creating a new diagnostic tool. Yet, as inscribed by design thinking, the most valuable design is when a solution is not only feasible (can we do this?), but also desired by the customer (do they want this?) and viable from a business perspective (should we do this?). As engineers, we are most comfortable—and quite good at—focusing on the feasibility lens. Yet, from an entrepreneurial perspective, we have to couple our inherent maker mentality with a constant awareness of business viability and the desire of the customer. Thus, as engineering educators, we similarly need to bring the business viability and the desire of the customer into our teaching so that we can truly and wholly focus on value creation.

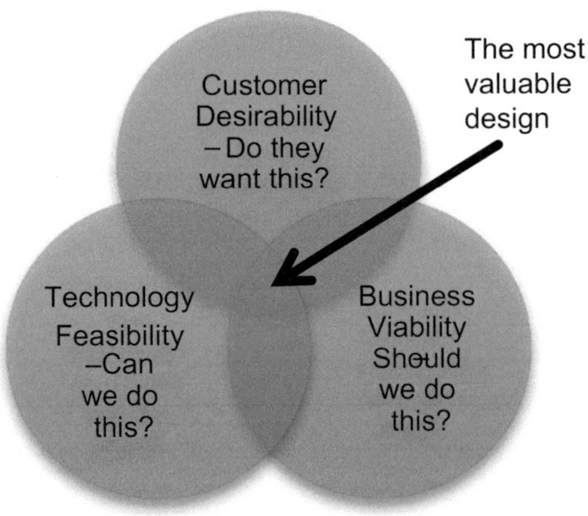

***Insight #5: Being entrepreneurial involves embracing uncertainty*.**

Entrepreneurial opportunities are, by definition, new and thus involve a high level of uncertainty. This is reinforced by Stevenson's recognized need for resourcefulness. This uncertainty not only relates to finding a way to build the solution (feasibility), but also uncertainty as to how the customer will react (desirability) or whether the solution is viable from a business sense (viability).

Uncertainty requires continuous experimentation, learning, and adapting. In his book, The Lean Start-up, Eric Reis refers to this process as the build-measure-learn loop. Indeed, the entrepreneurial process necessities perseverance in shifting from a state of assumption to knowledge. As engineering educators, our students are already familiar with the scientific method and the engineering design process—which are both the basis of the build-measure-learn loop. Yet, we need to again ensure that this process is being applied not only to the feasibility, but also to the customer desirability and business viability angles. We also need to stress the need to embrace uncertainty and learn to adapt accordingly.

2.3 Putting it Together: The Entrepreneurial Mindset

Entrepreneurial mindset: The inclination to discover, evaluate and exploit opportunities

Given our focus on the entrepreneurial mindset, we are thereby interested in the inclination towards entrepreneurship. In other words, the entrepreneurial mindset is ***the inclination to discover, evaluate, and exploit opportunities***.

In their book, entitled The Entrepreneurial Mindset, McGrath and MacMillan (2000) suggest that you know you have fully embraced the entrepreneurial mindset when you start to act and think like a *habitual entrepreneur*. As suggested by the name, habitual entrepreneurs are known for making a career out of starting businesses, both within existing organizations and as independent ventures. While you do not need to actually start businesses to embrace the entrepreneurial mindset, the point is that you utilize the entrepreneurial way of thinking so frequently that it has become a habit. McGrath and MacMillan (2000: 2–3) go on to identify five characteristics of the entrepreneurial mindset in common with habitual entrepreneurs:

1. They passionately seek new opportunities.
2. They pursue opportunities with enormous discipline.
3. They pursue only the very best opportunities and avoid exhausting themselves and their organizations by chasing after every option.
4. They focus on execution—specifically, adaptive execution.
5. They engage the energies of everyone in their domain.

If the entrepreneurial mindset involves an inclination to act or think in a certain way, can we, as engineering educators, actually impact how our students act or think? Can we really change their mindset? **Yes, but it requires practice so that it becomes a habit**. And, this practice must be grounded in theory. As outlined by Neck et al. (2014: 9):

> …in order to *learn* entrepreneurship, one must *do* entrepreneurship. Our position, however, is that doing entrepreneurship does not exclude theory. On the contrary, effective doing of entrepreneurship requires a set of practices and these practices are firmly grounded in theory.

References

Byers, T., Seelig, T., Sheppard, S., and Weilerstein, P. 2013. Entrepreneurship: Its role in engineering education. *The Bridge on Undergraduate Engineering Education* 43 (2): 35–40.

Drucker, P.F. 1985a. The Discipline of innovation. *Harvard Business Review*.

Drucker, P.F. 1985b. *Innovation and Entrepreneurship Practice and Principles*. New York: Harper & Row.

Dweck, C.S. 2006. *Mindset: The new psychology of success*. New York: Ballantine Books.

Gupta, A.K., and V. Govindrarajam. 2002. Cultivating a global mindset. *Academy of Management Executive* 16 (1): 116–126.

McGrath, R.G. and I. MacMillan. 2000. *The Entrepreneurial Mindset*. Harvard Business School Press.

Neck, H.M., P.G. Greene, and C.G. Brush. 2014. *Teaching Entrepreneurship: A Practice-based Approach*. Edward Elgar Pub.

Sarasvathy, Saras. 2005. *What makes entrepreneurs entrepreneurial?* Darden Business Publishing.

Schumpter, J.A. 1942. *Capitalism, Socialism and Democracy*. Harpers and Brothers.

Shane, S., and S. Venkataraman. 2000. The promise of entrepreneurship as a field of research. *Academy of Management Review* 25 (1): 217–226.

Stevenson, H. 1983. *A perspective on entrepreneurship*. Harvard Business School Publishing. Boston, MA.

Thum, Myrko. 2012. *The right mindset: Change your mindset in 6 Steps*. Retrieved from: http://www.myrkothum.com/mindset/.

Zappe, S., K. Hochstedt, and E. Kisenwether. 2013. Teaching to innovate: Beliefs and perceptions of instructors who teach entrepreneurship to engineering students. *International Journal of Engineering Education* 29 (1): 45–62.

Chapter 3
Why Is the Entrepreneurial Mindset Important to Future Engineers?

Author Reflection

Several years ago, I participated in an accelerator start-up program targeting technology ventures. Of the ten teams selected, the majority of the participants had an engineering or computer science background with limited education or experience in entrepreneurship or business concepts. The program connected each team with a business or entrepreneur mentor and required participants to learn about the lean start-up process through the eyes of the business model canvas. I walked away from that program with two major takeaways. First, I was blown away by the quantity of life changing information I was able to learn in such a short period of time. I now had knowledge of the skills and mindset necessary to start a company and/or add more value to my current employer. Second, I walked away from the program rather disappointed and perplexed wondering why we [higher education] are not sharing this information with engineering students. I wondered if we were doing our students a dis-service by keeping this information from them. From that point on I decided I was going to do whatever was possible in my control to share and embed the entrepreneurial mindset in all courses that I teach.

The purpose of this chapter is to explain the role of engineers in the entrepreneurial process, provide reasoning for how and why industry demands the entrepreneurial mindset in engineering graduates, and provide motivation for how engineering educators can benefit from incorporating the entrepreneurial mindset into the classroom, and ultimately engineering students.

L. Bosman and S. Fernhaber, *Teaching the Entrepreneurial Mindset to Engineers*, DOI 10.1007/978-3-319-61412-0_3

3.1 Industry Demands the Entrepreneurial Mindset in Our Graduates

Within existing corporations, much of the entrepreneurship and innovation occurs within dedicated research and development (R&D) efforts. The National Science Board's Science and Engineering Indicators 2016 Digest reveal a significant, and growing, reliance on engineers through research and development worldwide:

- R&D intensity is the proportion of a country's economic activity (gross domestic product, or GDP) devoted to R&D investment. China's R&D intensity has increased sharply over time, as growth in R&D outpaced a rapid expansion in GDP. China's R&D intensity now exceeds that of the EU, but it remains well below that of South Korea—which has also sharply increased its R&D intensity over time—and of the United States.
- Research and development (R&D) expenditures worldwide are estimated to have reached nearly $1.7 trillion in 2013, doubling from $836 billion a decade earlier.
- Global R&D expenditures are highly concentrated in three regions: East/Southeast and South Asia, North America, and Europe. The seven countries with the largest R&D expenditures together accounted for nearly three-fourths of total global R&D in 2013. The United States remains the largest R&D performer and accounted for 27% of total worldwide R&D in 2013. China is now the second largest R&D performing nation, accounting for 20% of the worldwide total.
- Asian countries have led the growth in worldwide R&D expenditures over the past decade, with China accounting for about one-third of the total global growth between 2003 and 2013. The United States and the European Union (EU) together accounted for approximately another one-third of the global growth during this period.

Innovation through R&D is important to maintaining global competitiveness, and for solving pressing problems. Yet, competitive growth involves more than the design or invention of a new product. It also requires being able to identify the market, understand the value proposition, create a viable business model, prototype and test, work with a team, source materials, communicate, lead, problem solve, persevere, and critically, deal with uncertainty. In short, it requires an entrepreneurial mindset.

It is no longer adequate for engineering graduates to be technically qualified; instead, to succeed, engineers must contribute to the success of the organization they join (Weilerstein and Byers 2016: 1). When hiring engineers, the top five skills sought after include (1) technical skills, (2) communication skills, (3) interpersonal skills, (4) problem solving and critical thinking, and (5) enthusiasm, commitment, and motivation (Brown 2016). The latter four are "professional skills" commonly associated with an entrepreneurial mindset.

In order to discover, evaluate, and exploit opportunities, individuals must work collaboratively in diverse environments—as diversity is not only proven to result in more creative ideas, but also descriptive of the setting in which most ideas are needed to be effective. Furthermore, entrepreneurial actions require a team. Even if the engineer is able to focus in on the product development role and bring in other team members to develop the product into a viable business, he/she still needs to understand the process and be able to work together for a common goal. Indeed, this ability to work with others, especially in other disciplines, requires empathy for their skill set and an understanding on how the different roles complement each other. Moreover, the ideas need to get buy-in. Engineers need to be able to communicate and rely on both their critical and creative thinking skills.

> The demand for the entrepreneurial mindset is not specific to a single engineering career path or discipline.

The demand for the entrepreneurial mindset is not specific to a single engineering career path or discipline. Rather, we would argue that by developing an entrepreneurial mindset you are able to better create value and leverage your expertise regardless of where it is applied (e.g., research, design, sales, consulting). This is consistent with the highly demanded "T-shaped" employee characterized by two kinds of characteristics; the expertise or depth of skill is represented in the vertical stroke of the "T," while the ability to connect with different areas of expertise is reflected in the horizontal stroke of the "T" (Hansen 2010).

3.2 The Entrepreneurial Mindset Is a Growing Movement

The demand for entrepreneurship education within engineering is growing rapidly by both students as well as by faculty that recognize the importance of doing so. These efforts have been reinforced in multiple ways. Formed in 2011, the mission of Stanford's Epicenter is to "empower U.S. undergraduate engineering students to bring their ideas to life for the benefit of our economy and society." As of 2015, the Epicenter has worked with 37 schools to help redesign their courses and programming. Their University Innovation Fellows program has worked with 607 students at 143 schools to become empowered as change agents.

Consisting of about 30 engineering schools as members, the Kern Entrepreneurial Engineering Network (KEEN) is a "national partnership of universities with the shared mission to graduate engineers with an entrepreneurial mindset so they can create personal, economic, and societal value through a lifetime of meaningful work." Their strategy is to develop peer collaboration and sharing of resources, and host an annual conference and variety of workshops. KEEN suggests that engineering curriculum should go beyond teaching the basic topics, and to instead stimulate curiosity and promote connections of classroom topics to real-world value creation opportunities.

Other networking or learning opportunities include the Midwest Engineering Entrepreneurship Network (MEEN) as well as a host of entrepreneurship education networks within the business arena. Pedagogical articles continue to increase, with the *Advances in Engineering Education* journal even putting out a special issue focusing specifically on entrepreneurship and innovation.

Engineering educators have the potential to make a difference when a student actually stumbles upon a validated design innovation. Many academic institutions are starting to increase the classroom focus on radical innovation and a multidisciplinary approach to implementing design projects in engineering courses (The Royal Academy of Engineering 2012). For the majority of student-developed design ideas that remain un-validated, and do not make it to market, the students still benefit. Whether it is student projects that have potential or a product developed post-graduation, we owe it to our students to help them develop their entrepreneurial mindset in order to empower them to leverage this thinking process and skill set. It will not only benefit the students, but also the customers that will appreciate the innovative solutions and also our economy as a whole.

Those engineering educators who fail to get on board will likely get left behind. In the book, "A Whole New Engineer," Goldberg and Somerville (2014) highlight the Olin College of Engineering (Needham, MA) and the University of Illinois iFoundry Incubator (Urbana, IL) on their successful approach to transforming engineering education with educational practices promoting the growth mindset by actively engaging thought processes including analytical, design, people, linguistic, body, and mindful. The revolution in engineering education is coming and engineering faculty need to be prepared to explore and promote new ways of teaching and learning.

Yet, while the movement towards integrating entrepreneurship into engineering education exists and is picking up steam, there is still a long way to go before every engineering student is reached. What we do know is that helping the individual faculty member is the most effective way to impact students, which ultimately is the purpose of this text. And at the end of the day, it is the engineering educator's responsibility to teach and encourage their students to convey innovative and entrepreneurial thinking (Byers et al. 2013).

3.3 How Engineering Educators Benefit by Integrating the Entrepreneurial Mindset

Entrepreneurship is exploding in relevancy and necessity. As educators, we have a responsibility to properly prepare our students so that they might make a difference and unleash their entrepreneurial spirit within their lives and careers. That being said, there are additional benefits for faculty who incorporate the entrepreneurial mindset into the engineering classroom.

Most notably, engineering educators benefit because of the increase in student motivation for learning the subject. In the book, "How Learning Works: 7

Research-Based Principles for Smart Teaching," Ambrose et al. (2010) suggest that one of the main factors motivating students to learn is whether or not students see value in the topic. Incorporating the entrepreneurial mindset into the engineering classroom is one way that allows students to see the value and understand the practical purpose of the topic and how it contributes to the economy. As a result, students might be less like to skip or sleep in class, and more likely to show an increased interest and understanding in the topic as demonstrated through assignments, class projects, and the exam.

Moreover, entrepreneurially minded engineering students make great student researchers and student research assistants. These students understand the customer's needs (which can be thought of as the needs of the engineering faculty research advisor) and characteristically strive to optimize productivity and quality in an attempt to meet the desired outcomes identified by the advisor. This is a particularly great quality for students assisting engineering faculty researchers participating in an NSF I-Corps project (a program that allows scientists and engineers the opportunity to explore the lean start-up process), where a student "entrepreneur" is typically allotted the brunt of the work. Entrepreneurially minded engineering students also make exceptional team members on federally funded Small Business Technology Transfer (STTR) and Small Business Innovation Research (SBIR) grant projects, which are essentially joint ventures between small businesses and universities aimed at accelerating research, innovation, and the commercialization process.

Lastly, teaching a new topic requires being a student first. As a result, engineering faculty that teach the entrepreneurial mindset will naturally increase their own entrepreneurial mindset. Entrepreneurially minded people have the ability to discover an opportunity, evaluate it, and exploit. So far, we have mentioned the benefit of this in relation to a business perspective. However, entrepreneurially minded faculty are capable of recognizing opportunity in all facets of their lives, from personal lives to research lives to teaching lives because there are "customers" all around us. There are always better, more strategic, and efficient ways to handle teaching, write and submit research papers for publications or grant proposals for funding, apply for teaching and research awards, more effectively schedule faculty responsibilities to allow for an increase in quality family time, etc. Entrepreneurially minded faculty excel at optimizing the work–life balance to fit the goals set for themselves.

References

Ambrose, S.A., M.W., Bridges, M., DePietro, M.C., Lovett, M.K., Normal, and R.E. Mayer. 2010. *How Learning Works: Seven Research-based Principles for Smart Teaching*. Jossey-Bass.

Brown, M. 2016. 5 Skills hiring managers look for in engineering grads. www.engineering.com

Byers, T., T. Seelig, S. Sheppard, and P. Weilerstein. 2013. Entrepreneurship: Its role in engineering education. *The Bridge on Undergraduate Engineering Education.* 43 (2): 35–40.

Goldberg, D.E. and M., Somerville. 2014. *A Whole New Engineer.* ThreeJoy Associates, Inc.

Hansen, M.T. 2010. IDEO CEO Tim Brown: T-shaped Starts: The Backbone of IDEO's Collaborative Culture. Retrieved from: http://chiefexecutive.net/ideo-ceo-tim-brown-t-shaped-stars-the-backbone-of-ideoae%E2%84%A2s-collaborative-culture/

National Science Board. 2016. Science & Engineering Indicators: 2016 Digest. Retrieved from: https://www.nsf.gov/nsb/publications/2016/nsb20162.pdf

Royal Academy of Engineering. 2012. Educating Engineers to Drive the Innovation Economy. London.

Weilerstein, P. and T., Byers 2016. Guest editorial: Entrepreneurship and innovation in engineering education. *Advances in Engineering Education.*

Part II
Integrating the Entrepreneurial Mindset into Existing Engineering Curriculum

Chapter 4
Being Intentional About Incorporating the Entrepreneurial Mindset into Engineering Curriculum

Author Reflection

Throughout the past several years, I have participated in faculty professional development workshops, where faculty come together to share best practices and learn from each other. Whenever the subject of active learning comes up, at least one brave soul (or "devil's advocate" faculty) provides push back suggesting that (a) students prefer to remain in their "passive learning comfort zone" and disengage when requested to activity participate in the classroom, resulting in poor teaching scores or (b) his or her particular course is content rich in that there is not enough time to incorporate active learning in the classroom, or even get through the high quantity of material already required for the course.

Sometimes faculty might feel that it is too risky to try new pedagogical approaches in the classroom and prefer to stick to traditional lecturing. This is understandable, as many institutions have a tendency to place a decent amount of weight on teaching evaluation scores. Also, for most untenured faculty, it is difficult to find a balance between research, teaching, and service.

That being said, modifying teaching strategies to incorporate the entrepreneurial mindset does not have to be hard and does not have to be time-consuming. Instead, a little prep work at the beginning of the semester can go a long way.

The purpose of this section is to provide an easy understanding of how to incorporate the entrepreneurial mindset into the engineering curriculum.

4.1 The Role of Engineers in the Entrepreneurial Process

Entrepreneurship is the engine of economic growth, and engineers play a clear role in substantial advances that enable this to occur.

From the macro-perspective, engineers work on projects from discovery to evaluation to exploitation, as shown in Fig. 4.1. From an engineer's perspective,

L. Bosman and S. Fernhaber, *Teaching the Entrepreneurial Mindset to Engineers*, DOI 10.1007/978-3-319-61412-0_4

Fig. 4.1 Opportunity Recognition Process

what starts as a simple new product or service idea during the discovery phase continues to get bigger as assumptions are evaluated, potentially snowballing into a business model opportunity to be exploited at the commercial level.

From the micro-perspective, the engineering design process (Fig. 4.2) can be considered as a series of iterative steps that engineers cycle through to design a solution to a customer problem or challenge. First, the problem is defined. The problem incorporates customer feedback, including their needs, wants, and desires, while considering general constraints commonly including resources such as manufacturing technology, time, and money. These objectives are then converted into design specifications to ensure the end product or service meets the needs specified by the customer. Second, brainstorming takes place. Brainstorming provides an opportunity to get all ideas, even the "unrealistic" ones, out in the open. This gives the design team thought space to think outside the box when design a solution to meet the needs of the customer. Third, the brainstorming is narrowed down to one or a few solutions and a plan is made to further evaluate the solution. Next, a prototype model is created and tested with the customer base. That feedback provides the design team and change to reflect and redesign a potential solution. It is important to note that the iterative aspect of the engineering design process allows the team to continuously consult with the customer to obtain feedback. The design team seeks to ensure the customer needs, wants, and desires are being met,

Fig. 4.2 Engineering Design Process

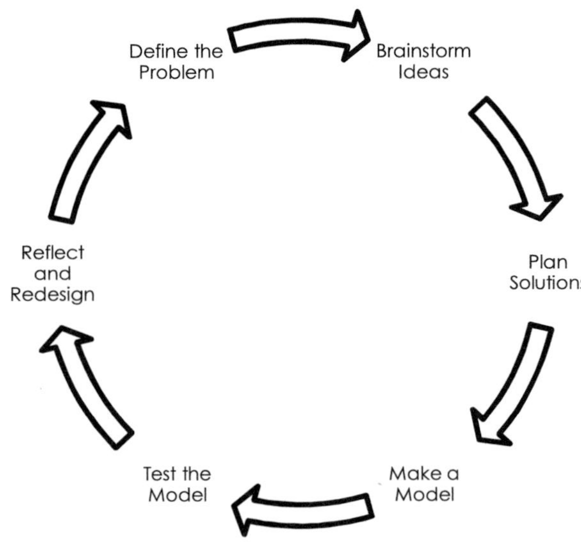

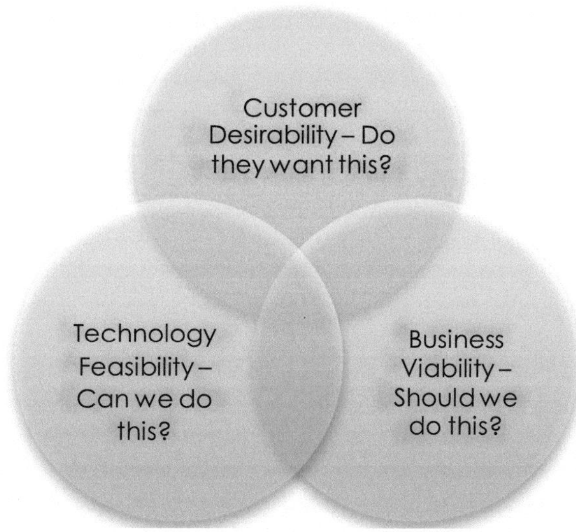

Fig. 4.3 Considerations for the Most Valuable Design

and at the same time, the design team seeks to ensure the company is capable of commercializing the product or service within the given technology, time, and money constraints.

All along, the engineering design team is striving to develop products and services that maximize design characteristics (Fig. 4.3) related to customer desirability, technology feasibility, and business viability.

Individually, a focus on customer desirability, technology feasibility, and business viability allows engineering design teams to validate questions including Do they want this? Can we do this? Should we do this? This is a starting point; however, individually validating these components is not enough. There are many innovations that customers might desire; however, they currently are not technology feasible or make business sense. For example, personally aircraft transportation vehicles or mind-reading applications. Eventually, these products might make it to market, but they are currently not available due to the lack of technology feasibility or business viability. A similar, more obvious analogy can be made for products and services that are technology feasible; however, lack of customer desirability or business viability, or products and services that are business viability, but are not desired by customers or technology feasible. The bigger issue in these cases is commonly the lack of customer desirability. If an innovation does not meet customer needs, the opportunity for technology feasibility and business viability ceases to be of much importance.

Therefore, innovations should, at a minimum, focus on an overlap between design characteristics that will further increase the likelihood of success. For example, functional innovations that focus heavily on Customer Desirability and Technology Feasibility can add value related to human–computer interactions, aesthetics and form, and design for sustainability. Process innovations that focus more so on Business Viability and Technology Feasibility can add value related to increasing productivity, increasing quality, and decreasing safety issues. Emotional innovations that focus deeply on Customer Desirability and Business Viability can add value related to marketing, branding, and relationship building.

Ultimately, optimizing these three characteristics (customer desirability, technology feasibility, and business viability) will create the most value toward the exploitation of an opportunity.

In summary (Fig. 4.4), as the design process moves through time from discovery to evaluation to exploitation, engineers continually iterate through the engineering design process with the ultimate goal to validate customer desirability, technologically feasibility, and business viability. However, none of this is possible without effective professional skills including collaboration and communication. Furthermore, practice and feedback allow habits to form into a mindset. In summary, the entrepreneurial mindset is important to future engineers because engineers provide the foundation for entrepreneurial quests and ultimately, economic growth.

As a result, faculty should be cognizant of four major intentions when integrating the entrepreneurial mindset into existing, engineering curriculum.

- ***Intention 1: The learning activity should provide an experience to discover, evaluate, and/or exploit opportunities. Opportunities that create the most value should be aimed at customer desirability, technology feasibility, and business viability***.
- ***Intention 2: The learning activity should provide an experience to develop professional skills (collaboration and communication)***.
- ***Intention 3: The learning activity should provide an experience for continued practice, reflection, and feedback***.
- ***Intention 4: The learning activity should be aligned with and reinforce the learning goals, learning objectives, and learning assessment***.

We next elaborate in more details in each of these intentions.

4.2 Intention 1: Curriculum Design for the Entrepreneurial Process

Intention 1: The learning activity should provide an experience to discover, evaluate, and/or exploit opportunities. Opportunities that create the most value should be aimed at customer desirability, technology feasibility, and business viability.

Entrepreneurial thinking is process oriented and happens over time. Thus, there needs to be a time variant aspect that considers a focus on discover, evaluation, and/or exploitation of an opportunity. Further, entrepreneurship is about value creation where the most valuable design opportunity is grounded at the overlap of customer desirability (Do they want this?), technology feasibility (Can we do this?),

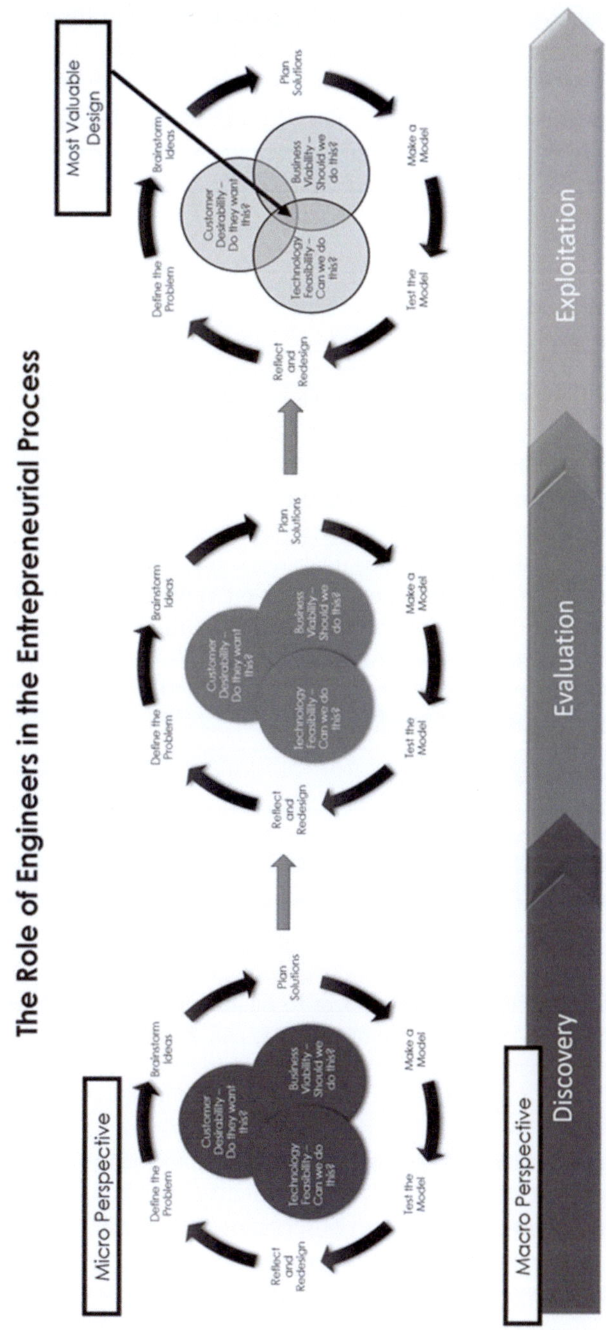

Fig. 4.4 The Role of Engineers in the Entrepreneurial Process

and business viability (Should we do this?). Thus, curriculum should be designed to investigate customer desirability, technology feasibility, and/or business viability.

Discovery

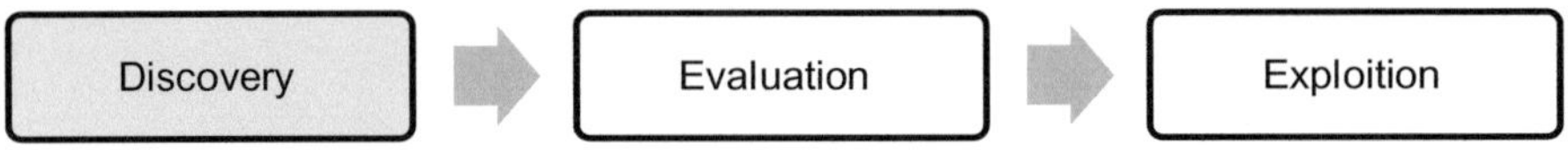

The discovery of value creation opportunities focuses on exploration and investigation of an opportunity with respect to customer desirability, technology feasibility, and business viability. Here, faculty can inspire students about promoting curiosity and questioning of the how, what, and why of currently accepted solutions. Students are encouraged to explore and consider alternate ways of doing things.

Qualitative research assignments offer one way to incorporate discovery into the engineering classroom. Example qualitative techniques include ethnographic, narrative, or phenomenological research (e.g., reading documents, watching videos, visiting places, field trip, conducting interviews, observation).

Discovery research questions should focus on value creation opportunities with respect to customer discovery, technology feasibility, and business viability. Sample research questions are provided in Table 4.1. These questions can be applied with respect to consideration of an innovation (product or service) that can be connected to the engineering class topic.

Evaluation

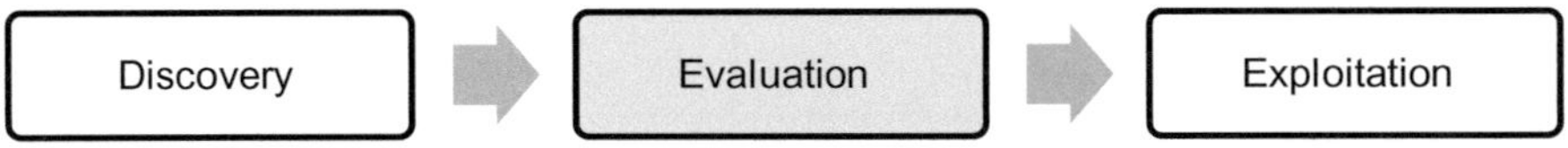

The evaluation of value creation opportunities focuses on the assessment and valuation of an opportunity with respect to customer desirability, technology feasibility, and business viability. Here, faculty can engage students involving the integration of information from multiple sources to further analyze an opportunity. Students are provided opportunities to assess risk and develop a plan for managing uncertainty.

Quantitative research assignments offer one way to incorporate evaluation into the engineering classroom. Example quantitative techniques include surveys, experimentation, secondary data, and observational research.

Evaluation research questions should focus on value creation opportunities with respect to customer discovery, technology feasibility, and business viability. Sample research questions are provided in Table 4.1. These questions can be applied with respect to consideration of an innovation (product or service) that can be connected to the engineering class topic.

Table 4.1 Example Research Questions Related to Entrepreneurial Opportunities

	Discovery Questions	Evaluation Questions	Exploitation Questions
Customer Desirability	What methods of exploration will be used to better understand and empathize with our customers? What might our customers want or need?	What assessment techniques will be applied to validate and refine the wants or needs of our customers?	What strategies will be applied to ensure customer wants or needs are met during implementation and beyond?
		Do they want this?	
Business Viability	What methods of exploration will be used to consider the financial implications? What might the revenue and cost structure be?	What assessment techniques will be applied to validate and refine the revenue and cost structure?	What strategies will be applied to optimize the revenue and cost structure to ensure a sufficient return on investment during implementation and beyond?
		Should we do this?	
Technology Feasibility	What methods of exploration will be used to design an initial solution? What might a prototype look like?	What assessment techniques will be applied to validate and refine the prototype, given the resources and capabilities available?	What strategies will be applied to leverage the necessary resources and capabilities to produce the designed solution during implementation and beyond?
		Can we do this?	

Exploitation

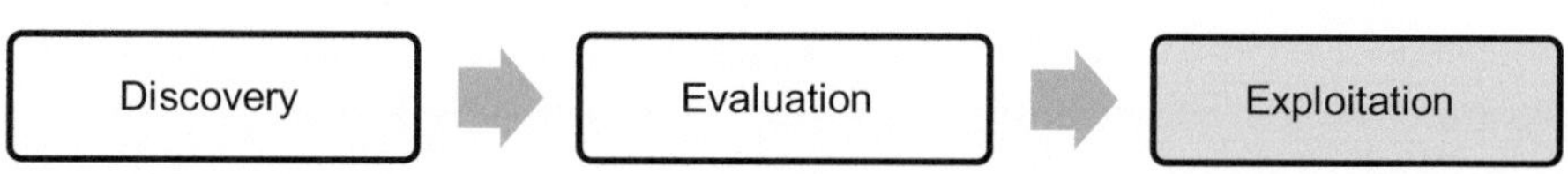

The exploitation of value creation opportunities focuses on promoting persistence in the design of an opportunity with respect to customer desirability, technology feasibility, and business viability. Here, faculty can use scaffold projects to encourage students to persist through and learn from failure. Students are guided through a formal or information design process culminating with an identification of a value creation opportunity.

Design opportunities offer one way to incorporate exploitation into the engineering classroom. Example design and valuation creation processes include the standard engineering design process, IDEO and d.School's Design Thinking, Google Design Sprint, and the Lean Startup Model.

Exploitation projects should focus on value creation opportunities with respect to customer discovery, technology feasibility, and business viability. Sample research questions are provided in Table 4.1. These questions can be applied with respect to consideration of an innovation (product or service) that can be connected to the engineering class topic.

4.3 Intention 2: Curriculum Design for Professional Skills

Intention 2: The learning activity should provide an experience to develop professional skills (collaboration and communication).

Innovation and design is very rarely done in isolation, which is why professional skills are critical components of the entrepreneurial mindset. In particular, developing collaboration and communication skills will provide students with the ability to work with peers on a design team, enhancing the likelihood of obtaining a quality solution.

Collaboration

Collaboration provides an opportunity for deeper learning, allowing students to engage in high-level tasks, discussion and debate, and shared decision-making. Collaboration provides students with the opportunity to form and work in teams, ultimately developing the ability to understand other team members' motivations and perspectives. This is of particular importance when working within a diverse team, as team diversity helps further promote the benefits of collaboration. Diverse design teams (with varying backgrounds, interests, and cultures) will produce better scientific, technological, and innovative results.

Diversity in teams is great; however, it is important to keep in mind legal statutes which prevent faculty from assigning students to teams on the basis of race, color, religion, creed, sex, sexual orientation, gender identity, national origin, ancestry, age, veteran status, disability unrelated to job requirements, genetic information, military service, or other protected status. An alternative is self-selection or randomly assigning students to teams, each of which have their own benefits.

Allowing students to self-select into teams can have many benefits. First, teams tend to stabilize quickly and efficiently, and perform more productively when students are empowered to choose who they work with. Second, self-selected teams tend to be happiest when working with people they choose to be around.

Random assignment of students into teams can also have many benefits. First, it is much more realistic of a workplace project in that employees typically do not get to choose their collaborators. Second, random assignment forces students to acquire and build additional professional skills such as developing trust in others and going through the steps of team development (e.g., forming, storming, norming, and performing). Going through this process will better prepare students for working with new people in a real-world context.

Communication

Practicing communication provides students with the opportunity to convey engineering design solutions in a variety of terms such as economic, manufacturing, and environmental, ultimately developing the ability to logically substantiate claims using data and analysis.

Communication, exchanging information, can occur in many different forms including oral, written, and visual, to name a few. There are many ways to practice communication, which can also be used as a means of assessment. Example opportunities are shown below.

- Oral: Speech, Debate, In-Class Discussion
- Writing: Online Discussion, Journal, Reflections, Essay, Minute Paper, Report
- Visual/Oral: Pitch Presentation, Poster Presentation, PowerPoint Presentation, Video, Product/Service Demonstration.

4.4 Intention 3: Curriculum Design to Promote Practice

Intention 3: The learning activity should provide an experience for continued practice, reflection, and feedback.

Practice molds habits into a mindset. Thus, the curriculum should allow room for multiple opportunities to apply entrepreneurial thinking for continued practice, reflection, and feedback.

Entrepreneurship is about creating value (e.g., discovery, evaluation, and exploitation of opportunities) under conditions of uncertainty. The greatest value is achieved when the work that engineers do is aligned, or balanced, with business viability as well as customer desirability. To teach entrepreneurship, we need to offer opportunities for students to PRACTICE this balancing act. By practicing (doing and learning), students can change the way they think and better respond to problem-solving under conditions of uncertainty. Essentially, we are helping them develop their effectual logic.

Entrepreneurship is a method requiring many iterations. As engineering educators, we can teach and promote the entrepreneurial mindset by providing multiple opportunities for students to encounter and practice key components of the entrepreneurial method.

Research suggests that instructional scaffolding, adding support and structure to a learning activity, helps students practice an activity that promotes increased learning outcomes. This can happen by giving students an opportunity to practice, reflect, and receive feedback during the planning and revision processes, including (but not limited to) idea generation, fine-tuning and refinement, evaluation, and writing. Throughout the different stages, feedback can happen with the help of peers, TA's, faculty, and/or the university writing center.

Here are a few more examples where we can practice the entrepreneurial mindset by considering the discovery, evaluation, and exploitation of opportunities (customer desirability, technology feasibility, and business viability).

First, students can practice coming up with topic-related ideas that are perceived to be desirable, feasible, and viable. This promotes the entrepreneurial mindset by requiring constant curiosity about the changing world, as it improves creativity. It also involves integrating information from many sources, as greater diversity in perspectives results in greater ideas. Yet, it also requires empathy for the customer, and initial thoughts of the revenue structure. Teamwork is essential in gathering diverse viewpoints, as is communication both with the customer and your team.

Second, students can practice evaluating ideas, and changing them, so that they become desirable, feasible, and viable. Contrary to assumptions, entrepreneurship is all about assessing and managing risk. For this reason, experimentation is essential. It helps ensure that you do not move forward until the idea has been validated. Teamwork is essential in gathering diverse viewpoints, as is communication both with the customer and your team.

Third, students can practice exploiting ideas so that they remain desirable, feasible, and viable. Here, students can develop collaboration and communication skills, which are necessary for completing most any business related projects.

4.5 Intention 4: Backward Course Design Process

Intention 4: The learning activity should be aligned with and reinforce the learning goals, learning objectives, and learning assessment.

To effectively develop the entrepreneurial mindset in our students, we must be purposeful in designing the curriculum. The Backward Course Design process (Wiggins and McTighe 2005), shown in Fig. 4.5, is an ideal approach for incorporating the entrepreneurial mindset into engineering courses. Here, the faculty member starts with broad learning goals (to incorporate the entrepreneurial mindset into the curriculum), then sets specific learning objectives, develops formative and summative assessments, and finally creates activities to support the learning objectives.

First, we start with the broad learning goal, which is to develop in our students the entrepreneurial mindset, defined as the inclination to discover, evaluate, and exploit opportunities. It is important to note that this would be a new learning goal in addition to the existing learning goals associated with the course content.

Second, we set specific learning objectives, which should be derived from the learning goal and associated requirements. Learning objectives should be context specific and customized to the course of study. Moreover, the learning objectives should be student focused; the learning objectives should be clear, breaking down the task with action verbs (e.g., Bloom's taxonomy) and conditions for performance; the learning objectives should be observable and measurable with criteria and standards for assessing the performance (Ambrose et al. 2010; Felder and Brent 2016; Nilson 2010).

Bloom's taxonomy (Fig. 4.6) is a framework for classifying educational goals using a continuum, whereas remember requires lower thinking skills and create requires higher thinking skills. Bloom's taxonomy is beneficial for faculty as a means to identify the level of thinking attributed to a particular assignment, module,

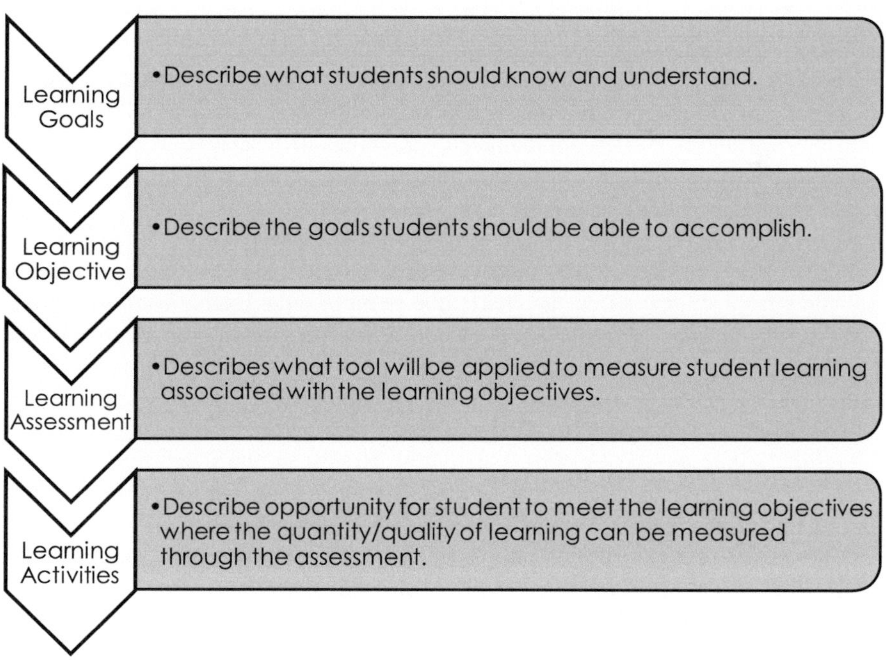

Fig. 4.5 Backward Course Design Process

Fig. 4.6 Bloom's Taxonomy

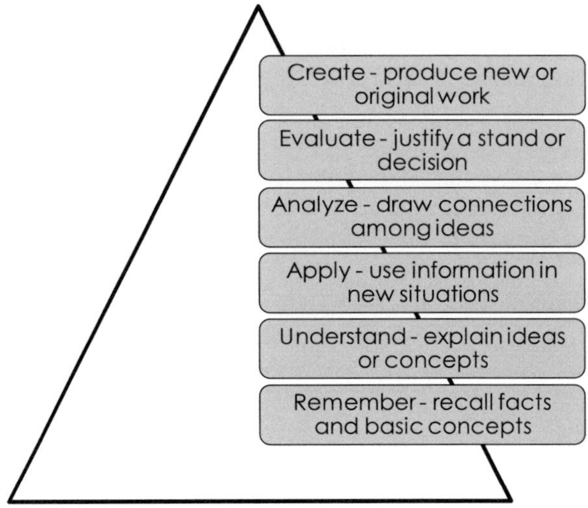

course, or program. It is recommended that a variety of educational goals to be applied to each type of assessment. For example, a homework assignment and exam should both require low-level thinking and high-level thinking, instead of a homework assignment only having low-level thinking and the exam only having high-level thinking.

Third, we develop formative and summative assessments. Formative assessments are part of the learning process and are used to assess student learning based upon the learning activities implemented. Examples of formative assessments can include homework, quizzes, class discussions, journals, and low-stakes assessment (e.g., minute paper, muddiest point, etc). The feedback from formative assessments provides a faculty insight into student needs for the purpose of making potential curriculum and learning activity modifications. Summative assessments are used to evaluate what a student knows at a particular point in time. Examples of summative assessments include state-mandated exams, placement tests or college entrance exams (e.g., SAT, ACT, Accuplacer, TOEFL), portfolios, and end of the semester exams. The feedback from summative assessments provides faculty and administration insight into learning goals, objectives, and program effectiveness.

Fourth, learning activities are created in alignment with the formative and summative assessments in an effort to support the learning objectives, and ultimate goal of developing the entrepreneurial mindset in students. Examples of learning activities can include lecture, readings, videos, and active learning activities. The next chapter highlights the use of online discussion prompts, value creation projects, and entrepreneurially minded tools.

4.6 Incorporating the Intentions into Online Discussions

Online discussions offer an ideal opportunity for incorporating the intentions into the engineering classroom, and provide many benefits for both students and faculty, alike. First, online discussions afford students the necessary time to provide a thought-provoking response and to consider other potential research or recent news media to support their responses. Second, online discussions provide students the opportunity to read and gain insight from other students' posts. Third, online discussions require limited class time and are an easy way to start incorporating the entrepreneurial mindset into the classroom. Finally, online discussions give educators the chance to provide immediate student feedback and ask further questions to dig deeper into the subject at hand.

Teaching Tips

Creating, deploying, and managing an entrepreneurially minded online discussion can be accomplished effectively and efficiently with a relatively minor investment in prep work prior to deploying the online discussion.

First, to effectively develop the entrepreneurial mindset in our students, we must be purposeful in designing the discussion prompts keeping in mind the framework previously mentioned.

Second, prior to deploying the discussion, it is recommended to set clear *student* expectations up front. Weekly asynchronous discussions are ideal because they allow students to participate during a time that is convenient for them. Students need to know up front how often they are expected to participate. Additionally, students should be provided a rubric to understand what signifies a quality discussion response.

Finally, prior to deploying the online discussion, it is recommended to set clear expectations up front. The faculty member should indicate how often he/she plans to participate, including times during the day and days during the week. For example, the faculty member could tell the students up front that he/she only plans to participate Monday–Friday between the hours of 8 and 10 am. If the online discussion has a large quantity of participants, it may be beneficial to assign student leaders to facilitate the discussion.

Potential Learning Activity Approaches

Example discussion prompt assignments are offered in Part 4. Several different approaches are demonstrated; however, all the examples provided were deployed in an online discussion format.

One approach requires responding to questions based on a specific reading, in-class speaker, or ethical dilemma. Using this approach, faculty members can either require answering a basic standard set of questions or customizing the questions taking into consideration the context.

Another approach requires the response to include technical references (typically, for a higher level course) or internet searches (typically, for a lower level course). Using this approach, faculty members can either use an open-exploration prompt ("find something you are curious about which relates to this week's topic") or guided-exploration prompt (which provides more customized and specific questions to guide the search).

Additional learning activity deployment options include face-to-face discussions, small-group student-lead presentations, journal writing, blog posts, essay, or research papers, to name a few. Also, prompts can be used as an end-of-class low-stakes assessment in the format of a 1-page summary, minute paper, or freewrite.

Example Prompts to Get Started

Identify a product or service related to TOPIC X. Identify the various stakeholders involved in the development, deployment, and use of the product or service.

Find out what stakeholders appear to like and dislike about the current product or service related to TOPIC X. Take into consideration aspects around customer desirability, technology feasibility, and business viability. Use one of the following data collection methods: (a) Observation, (b) Interview, (c) Survey, and (d) Trends in the media.

Given a product or service related to TOPIC X. Compare and contrast the currently available substitutes. What are the pros and cons of each? What improvements (and why) can be made to each with respect to customer desirability, technology feasibility, and business viability?

Given ARTICLE X, VIDEO Y, and/or TEXTBOOK READING Z related to TOPIC X, formulate an opinion on the state of a product or service.

4.7 Incorporating the Intentions into Value Creation Projects

Semester-long value creation projects provide another opportunity for incorporating the intentions into the engineering classroom, and provide many benefits for future engineers. First, engineers need to design new products and services with the value proposition and user needs in mind, and not simply based on technical and functional concepts taught in the traditional engineering classroom. Value creation projects allow for this alternative method of thinking. Second, value creation projects focus on the customer from the very beginning, allowing students to better understand the role of empathy in the design process. Third, value creation projects promote a method of solution-focused thinking, which allows students to think outside the box and apply active learning and creative thinking to theoretical concepts. Fourth, value creation projects allow for increased development of professional skills, including collaboration and communication, through group work culminating with a final presentation. Fifth, value creation projects increase motivation for learning by allowing students to see value by connecting real-world applications to the class topic. Furthermore, value creation projects are commonly completed outside the classroom, requiring limited class time. Finally, incorporating value creation projects into non-design engineering courses allows the engineering student to gain real-world engineering skills outside of the traditional and typical freshman and senior engineering design courses. Remember, developing an entrepreneurial mindset takes practice.

Teaching Tips

Value creation projects connect the traditional scientific method and engineering design process to business and marketing through a focus on goals rather than problems. This iterative process encourages engineering students to think outside the box, beyond the current state of parameters to consider the potential that the future holds. Creating, deploying, and managing entrepreneurially minded value creation projects can be accomplished effectively and efficiently with a relatively minor investment in prep work prior to deploying the projects.

First, the value creation projects should be developed keeping in mind the framework previously mentioned. Second, prior to deploying the value creation projects, it is recommended to consider how much time the faculty member has available to dedicate to project consulting and grading, in addition to the quantity of time students should be expected to work on the project. Faculty should consider using a grading rubric for the report and/or presentation, peer reviews to assist with student feedback, and requiring working with campus resources for writing assistance (if available).

Potential Learning Activity Approaches

Example value creation projects are offered in Part 4. Several different approaches are demonstrated.

Some projects are multi-part lasting the length of the entire semester, or even across two semesters, while other projects are packaged into a module last 2–3 weeks. However, for all projects, deliverables and assignments were required on a weekly or bi-weekly basis.

For many projects, the faculty member provided constraints through solving a very specific design challenge. However, several projects incorporated industry-specific challenges or allowed students to design their own project and constraints.

In many cases, the projects utilize the entrepreneurially minded tools highlighted in Chap. 10. The Design Thinking process offers a methodology combining an analytical and creative approach, drawing on innovation tactics grounded from the engineering and design paradigms. The Value Proposition Canvas offers a simple and visual way to understand and communicate the customer needs (what they want and don't want) and design product/service features (identifying how the design will meet customer needs). The Business Model Canvas offers a simple and visual way to describe, evaluate, and pivot the business model taking into consideration nine main components: Value Proposition, Customer Segments, Customer Relationships, Channels, Revenue Streams, Cost Structure, Key Resources, Key Partners, and Key Activities. Steve Blank's "How to Build a Startup" is offered for free through Udacity, a massive open online course, and is a great compliment to a flipped classroom (where the videos can be watched outside the classroom allowing for active learning inside the classroom to bring an idea from a conceptual design to market).

Project deliverables often included both a report and presentation. However, other assessment options can include one or a few of these components: poster presentation, infographic development using a free online tool such as Piktochart, narrated PowerPoint presentation (useful when class time is limited), and other visual presentation formats using online tools such as Canva or Sway.

References

Ambrose, S.A., M.W., Bridges, M., DiPietro, M.C., Lovett, and M.K., Norman. 2010. *How learning works: Seven research-based principles for smart teaching*: John Wiley & Sons.

Felder, R.M., and Brent, R. 2016. *Teaching and Learning STEM: A Practical Guide*: Wiley.

Nilson, L.B. 2010. *Teaching at its best: A research-based resource for college instructors*: John Wiley & Sons.

Wiggins, G.P. and J. McTighe. 2005. *Understanding by design, Ascd.*

Chapter 5
Abet Student Outcomes and the Entrepreneurial Mindset

Author Reflection

While finishing up grad school, I served as a Teaching Assistant (TA) for an engineering course. It was here that I received my first real introduction and education on the ABET accreditation process. Due to last minute requirements and a change in faculty assignments, I was asked to get involved in the accreditation process.

Toward the end of the semester, I was notified I needed to create a package that included a syllabus, textbook, assessment documents (e.g., assignments, tests, projects, homework, labs, etc.…) and sample student work including both high- and low-quality work. Here's the thing, like the effective and efficient TA that I aimed to be, I was on top of my game and already returned the assessment documents back to the student. Now, I had to request students to give everything back, being sure to get enough sample student work of varying quality.

In the whole scheme of things, my role in the accreditation process was minimal. However, any time someone mentions ABET, it is typically followed with a long sigh because we all have some understanding of the time and paperwork commitment required to keep up with ABET. That being said—if it were easy, everyone would do it. Keeping up with ABET accreditation assures engineering programs meet the agreed upon standards (developed by industry and academia) in an effort to produce engineering graduates capable of becoming leaders in innovation, emerging technologies and meeting societal needs.

The purpose of this chapter is to understand the perspective that ABET takes toward the entrepreneurial mindset and what that means for us, as engineering educators.

L. Bosman and S. Fernhaber, *Teaching the Entrepreneurial Mindset to Engineers*, DOI 10.1007/978-3-319-61412-0_5

5.1 Overview of ABET Student Outcomes

ABET, formerly the acronym for the Accreditation Board for Engineering and Technology which has now broadened its scope to include Computer Science and Applied Science in addition to Engineering and Technology, offers a grueling, yet highly sought after accreditation, for colleges or universities producing graduates in engineering or engineering technology at the associate, bachelor, or master degree levels. ABET accreditation provides assurance that programs are delivered in a way to develop graduates who are qualified and ready to enter technical engineering fields as leaders in innovation, taking into consideration ethical concerns for public welfare. ABET currently accredits 3709 programs offered at 752 colleges and universities spread-out over 30 countries.

ABET documents criteria for accrediting engineering programs on an annual basis utilizing a global network of experts from higher education, government, and industry. Collectively and over time, experts have developed a series of criteria relating to:

(1) Students
(2) Program Educational Objectives
(3) Student Outcomes
(4) Continuous Improvement
(5) Curriculum
(6) Faculty
(7) Facilities
(8) Institutional Support.

Student Outcomes, commonly referred to as the "(a) through (k)", state what graduating students are expected to know and do upon completion of the engineering program. ABET's 2016–2017 Criteria for Accrediting Engineering Programs (ABET 2016) provides general criteria for bachelor level programs, including a focus on student outcomes as follows.

Student outcomes are outcomes listed below plus any additional outcomes that may be articulated by the program:

(a) an ability to apply knowledge of mathematics, science, and engineering
(b) an ability to design and conduct experiments, as well as to analyze and interpret data
(c) an ability to design a system, component, or process to meet desired needs within realistic constraints such as economic, environmental, social, political, ethical, health and safety, manufacturability, and sustainability
(d) an ability to function on multidisciplinary teams
(e) an ability to identify, formulate, and solve engineering problems
(f) an understanding of professional and ethical responsibility
(g) an ability to communicate effectively

(h) the broad education necessary to understand the impact of engineering solutions in a global, economic, environmental, and societal context
(i) a recognition of the need for, and an ability to engage in life-long learning
(j) a knowledge of contemporary issues
(k) an ability to use the techniques, skills, and modern engineering tools necessary for engineering practice.

We posit that the ABET student outcomes can be categorized into four main areas including (1) General Education, (2) Engineering, Math, and Science Fundamentals, (3) Professional Skills, and (4) Engineering Design and Value Creation, as shown below.

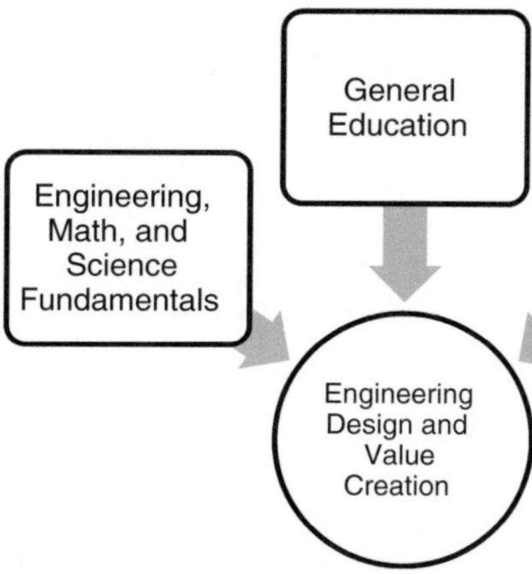

Engineering, Math, and Science Fundamentals

ABET criteria requires a minimum of 2.5 years (80 credits) worth of college-level engineering, math, and science courses. As such, the bulk of the engineering bachelor's degree will be grounded in courses focusing on engineering, math, and science. This provides an opportunity for engineering programs to support ABET's learning outcomes of (a), (b), (e), and (k):

- (a) an ability to apply knowledge of mathematics, science, and engineering
- (b) an ability to design and conduct experiments, as well as analyze and interpret data
- (e) an ability to identify, formulate, and solve engineering problems
- (k) an ability to use the techniques, skills, and modern engineering tools necessary for engineering practice.

General Education

ABET criteria requires a "general education component that complements the technical content of the curriculum and is consistent with the program and institution objectives." This often includes courses taught outside the college of engineering, such as the arts, humanities, or social sciences, to name a few. This provides an opportunity for engineering programs to support ABET's learning outcomes of (f), (h), and (j):

- (f) an understanding of professional and ethical responsibility
- (h) the broad education necessary to understand the impact of engineering solutions in a global, economic, environmental, and societal context
- (j) a knowledge of contemporary issues.

Engineering Design and Value Creation

ABET criteria requires that programs offer a major culminating engineering design experience based on the attainment of knowledge and skills obtained from earlier coursework. For the majority of engineering bachelor's programs this is commonly referred to as a senior design course. Here, students get the opportunity to take everything they have learned and apply it to a "real-world" design project. This provides an opportunity for engineering programs to meet ABET's learning outcomes of (c) and (i):

- (c) an ability to design a system, component, or process to meet desired needs within realistic constraints such as economic, environmental, social, political, ethical, health and safety, manufacturability, and sustainability
- (i) a recognition of the need for, and ability to engage in life-long learning.

Professional Skills

Unlike the previous three categories of required courses and experiences, ABET does not require stand-alone courses for professional skill development. Instead, the expectation is for collaboration and communication to be encouraged and developed throughout the engineering program including courses related to (1) General Education, (2) Engineering, Math, and Science Fundamentals, and (3) Engineering Design and Value Creation. This provides an opportunity for engineering programs to meet ABET's learning outcomes of (d) and (g):

- (d) an ability to function on multidisciplinary teams
- (g) an ability to communicate effectively.

5.2 Summary of the Entrepreneurial Mindset and ABET

What does this mean for engineering educators and the entrepreneurial mindset? One can easily make the case that the major design experience (e.g., senior design), by ABET definition, illustrates an example where students are given the opportunity to develop the entrepreneurial mindset [The inclination to discover, evaluate, and exploit opportunities]. Specifically, it highlights the technological feasibility, customer desirability and business viability, which are essential to simultaneously addressing when aiming for value creating opportunities.

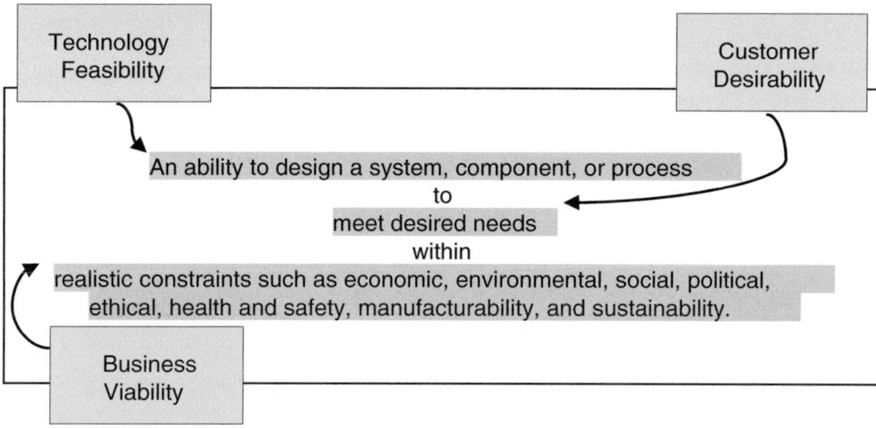

Yet, an entrepreneurial mindset cannot be effectively developed in the final experience in the last year. Instead, it requires practice and habit nurtured throughout the program, which we suggest could be incorporated into the other engineering courses.

General Education, Math, and Science courses are typically not offered through the college of engineering, and are therefore, out of the scope of influence for engineering faculty. Thus, engineering courses provide an ideal location for faculty to incorporate the entrepreneurial mindset.

In summary, incorporation of entrepreneurial minded experiences has the potential to meet many student learning outcomes required by ABET (the accreditation agency supported by the majority of engineering colleges). However, to effectively instill the entrepreneurial mindset in our students, we need more practice which can be achieved by integrating the entrepreneurial mindset into engineering courses early on and throughout the curriculum.

Part III
Where Do We Go From Here?

Chapter 6
Integrating the Entrepreneurial Mindset as an Engineering Educator

Author Reflection

Many assistant professors, within engineering and other disciplines, go straight into academics without the opportunity to practice engineering in 'real-world' applications. Many of these faculty are excellent researchers, however, due to the lack of experience may have a limited understanding of how today's business actually operates with respect to validating customer desirability, technology feasibility, and business viability. Thus, the question then becomes where they can go to attain this insight to help prepare future engineers. It is important that students see value in classroom topics and are able to connect theory and textbooks to real-world applications.

The purpose of this section is to provide faculty motivation for being entrepreneurial in our teaching approach and at the institutional level.

6.1 Being Entrepreneurial in Our Teaching Approach

We recognize that the thrust of this book has been a call to integrate the entrepreneurial mindset within engineering students. Yet, this can only happen if we, as engineering educators, further develop and depend on our own entrepreneurial mindset in the process. No, this does not mean that you need to go out and start your own business (that is, unless you want to). Rather we need to become more entrepreneurial as engineering educators. This involves being innovative in our teaching approach, being proactive or alert to new pedagogical opportunities, and lastly, take some risks in doing so.

First, we can innovate. A lot. Innovations may be incremental additions to existing courses, such as trying out new experiential exercises, incorporating innovation prompts or adding a new case. In other instances, we might have more freedom to innovate and develop new projects that span the entire semester. In

L. Bosman and S. Fernhaber, *Teaching the Entrepreneurial Mindset to Engineers*, DOI 10.1007/978-3-319-61412-0_6

however we innovate, we need to strive to develop new ways to engage the student. In many cases, this involves bringing the real world into the classroom. As evidenced by the learning pyramid, this also requires experiential learning (and lecturing less). It is a huge responsibility to innovate. While innovation can require a lot of time and energy, it is also a lot of fun and can be very rewarding when successful.

> We recognize that the thrust of this book has been a call to advance entrepreneurship education within engineering students. Yet, this can only happen if we, as educators, further develop and depend on our own entrepreneurial mindset in the process.

Second, we can be proactive and try to anticipate what pedagogical opportunities might exist. Thus, we need to stay very active in what is happening in the 'real world' and talk to 'real' entrepreneurs and community members. Being constantly on the look out for new teaching ideas, these may appear through various email list serves, conferences, scanning the web, reading the newspaper/magazines or talking with folks in the community. It means becoming active in local networks, such as KEEN, VentureWell or Epicenter. As we teach a class, we can keep a running list of ways in which we might want to explore improving upon the following semester. Thus, it is our proactiveness or alertness that feeds the innovations we pursue.

Third, we can take risks. This means that sometimes our course innovations work great, while other times they do not. Yet, as we teach our students, we do not necessary fail. Rather, we learn a bit more each semester on what works and what does not work. While entrepreneurs are risk-takers, they are moderate or calculated risk-takers. This means that they try to manage the risk by conducting research and/or spreading out the risk through collaboration with other partners. When creating a new assignment or exercise, we can similarly try to reduce the risk by sourcing proven ideas that have worked elsewhere at other institutions.

Our guess is that you already do a lot of this as it relates to the engineering disciplines in which you specialize. The next step is extending it to integrating the entrepreneurial mindset so that we can ensure our students develop the habits they need to best leverage their engineering skillsets. While the curriculum examples offered in the preceding section may be a starting point, a commonly asked question is "where do we go from here?" How do we get linked in with others that are implementing the entrepreneurial mindset? What resources exist that we can tap into? Fortunately, there are many resources that do exist. The first pool of resources that focuses exclusively on applying the entrepreneurial mindset within the context of engineering is small, but growing and gaining speed. Yet, a second pool of resources also exists. Although this second pool has been created from the point of view of the business school, there is a lot to be leveraged and that can be creatively applied. Thus, we hereby draw on both "pools" of resources to showcase how you, as the engineering educator, can leverage them.

6.2 Tips for Continued Development

When developing curriculum incorporating the entrepreneurial mindset, there are a few tips that can be taken to ensure that the process is both efficient and effective.

- ***Benchmark other institutions***. It is always useful to understand how entrepreneurship is being integrated into other courses. One way is to examine model institutions that have received some sort of acclaim or recognition for their entrepreneurship program. Within the entrepreneurship arena, for example, you might look at the model undergraduate programs in entrepreneurship as awarded through the United States Association of Small Business and Entrepreneurship (USASBE). Another option is to profile the member institutions of either the KEEN or Epicenter organizations. Keep in mind that the role of benchmarking is to see what is out there, and learn from these institutions. There may be some elements that you want to copy and take back to your institution, and still others that you might take the concept and customize it. Moreover, you might end up innovating and creating something altogether different.
- ***Leverage insight from local entrepreneurs (or, in this case, entrepreneurial engineers)***. Talk to engineers in a variety of contexts, whether it be a start-up, large corporation, growing firm, etc., and find out what skills they rely on.
- ***Leverage insight from employers***. Talk to your employers and again, ask they what skills they are looking for. Sometimes there can still be confusion as to what is meant by the entrepreneurial mindset, so be sure to focus on the components of the mindset and don't get stuck on terminology. See if your experience with employers matches what we read about. See which components of the entrepreneurial mindset seem to be lacking in your graduates, and where you might focus your efforts.
- ***Leverage insight from interdisciplinary faculty***. You cannot simply jump to the end point, but you can accelerate it, especially with partnerships across colleges. If we look at the history of how formal entrepreneurship programs develop, they typically start within the business school. Ironically, many of the design elements are simply being transferred from the engineering folks to the business folks—thus, there is already a lot of overlap. The key is to find these synergies.

The curriculum is just one way in which the entrepreneurial mindset can be taught to engineering students. There are also many programming opportunities and strategic partnerships within the university and greater community that can be leveraged. Many of these programming examples that are specific to your university size, location and focus can be identified by looking to model, peer, aspirant, or competitive institutions. However, we continue to reinforce that while it is important to examine these other programs, we do not necessarily recommend that you copy them. Some might make sense for you to copy, while others might make more sense for you to customize. And still, there is likely room for you to innovate all together in envisioning what entrepreneurial engineering could look like at your institution.

Chapter 7
Pedagogical and Professional Development Resources

Author Reflection

One of my colleagues recently offered a seminar on his year-long sabbatical where he had the opportunity to work for a year as an entry-level mechanical engineer for a globally recognized manufacturing company focused on industrial automation. During his presentation, we got to hear him talk about what courses [in his opinion] truly benefit and prepare graduating students for industry, what potential research and innovation opportunities exist for collaboration between industry and academics, the role engineers perform throughout the company (e.g., Sales Engineer, Design Engineer, Production Engineer, Project Management, Research and Development, Consulting, Facilities Manager, etc.…), and the latest communication and collaboration tools. His talk was very insightful. Similar to many engineering professors, I worked in industry [as a manufacturing engineer] prior to switching over to academics. However, it has been at least 10 years. My colleague's presentation allowed me to reflect on that notion that sometimes I find myself getting caught up in my academic role, losing sight of the ability to apply real-world experiences and storytelling in the classroom. For those of you, like me, who are looking for new ideas on making the engineering classroom more practical and are willing to do a little extra work outside the classroom, this chapter is for you.

7.1 Books

If you search on Amazon for books with the topics of entrepreneurship and/or engineering, you will actually find quite a few books. This is exciting, and below we offer our "Top" entrepreneurial book picks for engineers—which could be passed along to students and/or read by you, the engineering educator, to get up to speed on what it means to be entrepreneurial.

L. Bosman and S. Fernhaber, *Teaching the Entrepreneurial Mindset to Engineers*, DOI 10.1007/978-3-319-61412-0_7

Top Entrepreneurial Book Recommendations for Engineers

1. The Lean Startup by Eric Reis (2011)
2. Value Proposition Design: How to Create Products and Services Customers Want by Osterwalder, et al. (2014)
3. Business Model Generation: A Handbook for Visionaries, Game Changers, and Challengers by Osterwalder and Pigneur (2010)
4. A Whole New Engineer by Goldberg and Somerville (2014)
5. The Innovator's Dilemma by James Christensen (1997)
6. Made to Stick: Why Some Ideas Survive and Others Die by Chip & Dan Heath (2007)
7. The E-Myth Revisited by Michael E. Gerber (2004)
8. The Four Steps to the Epiphany by Steve Gary Blank (2013)
9. Zero to one: Notes on Startups, or How to Build the Future, by Peter Thiel (2014)
10. Design Methods: A Structured Approach for Driving Innovation in Your Organization (2012)

Another book that comes highly recommended, but it geared towards helping providing examples of class exercises that can be integrated into the classroom is "Teaching Entrepreneurship: A Practice Based Approach" by Neck et al (2014).

7.2 Journals

In addition to books, there are a number of journals that have emerged with a focus on aiding entrepreneurship educators. The materials offered range from short, entrepreneurial exercises to course wide projects and beyond. They serve as a useful database to peruse and also contribute to.

- ***Journal of Engineering Entrepreneurship***: This international, peer-reviewed journal is located at http://www.jeenonline.com, and is considered the main source for discussion and dialog between engineering entrepreneurship educators, practitioners, and researchers. The journal started in 2010 and publishes quarterly. It offers authors an opportunity to publish entrepreneurial engineering related research findings and teaching methods.
- ***Journal of Engineering Education***: Within origins dating back to 1893, its current mission is to "cultivate, disseminate, and archive scholarly research in engineering education." It can be found online at https://www.asee.org/papers-and-publications/publications/jee. While its main focus is not on the intersection of entrepreneurship and engineering, there are many articles that do focus on this topic.
- ***Advances in Engineering Education***: This online, peer-reviewed journal was created to compliment the *Journal of Engineering Education*, with the mission of "to disseminate significant, proven innovations in engineering education practice, especially those that are best presented through the creative use of

multimedia." Again, the main focus is not on the entrepreneurial mindset, but there are multiple articles that do emerge in this journal based on this topic. The journal is located at http://advances.asee.org/.

- ***International Journal of Engineering Education***: This international, peer-reviewed journal publishes six issues per year, has been around for about 30 years, and is located at https://www.ijee.ie/. The main focus is not the entrepreneurial mindset, however, from time to time, it solicits special issues related to specific engineering education topics.
- ***American Journal of Engineering Education***: This open access journal is located at https://www.cluteinstitute.com/journals/american-journal-of-engineering-education-ajee/. It publishes articles that connect theory and practice, cross-institution and cross-disciplinary innovation, and focus on educational strategies which consider the changing role of engineering in a local, global, and societal context. AJEE is particularly interested in articles promoting skills related to critical thinking and problem-solving, creativity, and a commitment to lifelong learning (which are all well-aligned with the entrepreneurial mindset).
- ***Journal of Entrepreneurship Education***: This double blind, peer-reviewed journal is located at http://www.alliedacademies.org/entrepreneurship-education/, and publishes empirical and theoretical research focused on entrepreneurship education and training. The journal started in 1999, first publishing annually and now publishes twice a year.
- ***Experiential Entrepreneurship Exercises Journal***: This peer-reviewed journal is located at http://launchideas.org/, and provides a collection of exercises to engage students in applying and practicing the entrepreneurial mindset. The journal started in 2014 and publishes up to four times per year. It enables and promotes more active entrepreneurial classrooms by providing a repository for sharing real world and tested exercises.

7.3 Workshops and Conferences

In recent years, the number of professional development workshops has increased. Below, we provide a mixture of workshops that specifically target entrepreneurial education (sometimes from an engineering viewpoint) as well as workshops that help advance entrepreneurial tools that are useful in the classroom.

- ***Kern Entrepreneurial Engineering Network (KEEN) Annual Conference***: KEEN consists of about 30 universities with the combined mission to graduate engineers with the entrepreneurial mindset. An Annual KEEN Conference allows engineering faculty the opportunity to share best practices and learn from each other for curriculum enhanced by incorporating the entrepreneurial mindset (http://engineeringunleashed.com/keen/).

- ***American Society of Engineering Education (ASEE) Annual Conference***: The Entrepreneurship and Engineering Innovation Division (ENT) at this annual conference can be especially helpful for both sharing your research as well as learning what new pedagogies you might be able to adapt (https://www.asee.org/).
- ***Lawrence Technical University ICE (Innovating Curriculum with the Entrepreneurial mindset) Workshops***: These are 3-day professional development events showcasing the important aspects of Entrepreneurially Minded Learning (EML), Active and Collaborative Learning (ACL), and Problem/Project Based Learning (PBL) and how to integrate them into engineering courses (http://engineeringunleashed.com/keen/innovating-curriculum-with-entrepreneurial-mindset-ice-workshops/).
- ***American Society of Engineering Education (ASEE) Annual Conference***: The Entrepreneurship and Engineering Innovation Division (ENT) at this annual conference can be especially helpful for both sharing your research as well as learning what new pedagogies you might be able to adapt (https://www.asee.org/).
- ***Villanova University—Deep Dive Workshop***: Villanova's Deep Dive workshop provides an immersive experiential learning experience that introduces faculty to engineering entrepreneurship principles and teaches them how to confidently integrate the entrepreneurial mindset into courses. Faculty from KEEN member schools is invited (http://engineeringunleashed.com/keen/june-13-17-villanovas-deep-dive-workshop/).
- ***Worchester Polytechnic Institute—Project-Based Learning Workshop***: Through collaborative work, teams of faculty and administrators at the Institute develop strategies to integrate project-based learning into their own undergraduate curricula, whether in general education or in the major, in one department or across the campus (http://wp.wpi.edu/projectbasedlearning/2017-institute/what-to-expect/).
- ***Olin Collaboratory Summer Institutes—Designing Student-Centered Learning Experiences***: Offered through Olin College of Engineering, participants from around the country around the world and provided the opportunity to conceive and catalyze change in engineering education (http://www.olin.edu/collaborate/collaboratory/summer-institute/).
- ***University of Florida—Experiential Classroom***: The Experiential Classroom is the premier clinic for faculty who are relatively new to the teaching of entrepreneurship and the building of entrepreneurship programs. This intensive, 3-day clinic is geared towards faculty members from any discipline who are incorporating entrepreneurship (http://warrington.ufl.edu/centers/cei/faculty/experiential/).
- ***Rose Hulman Institute of Technology—Making Academic Change Happen (MACH)***: MACH is an experience, a philosophy, and an approach. Most simply, MACH is a 3-day workshop during which change agents learn the skills they need to implement changes on their campuses. This approach harnesses the lessons of the research literature in institutional change and translates them into actionable

skills any faculty member can learn and utilize. This is a 3-day conference (http://www.rose-hulman.edu/offices-and-services/making-academic-change-happen.aspx).

- ***Price-Babson Symposium for Entrepreneurship Educators Program***: The Price-Babson SEE program is a weeklong program designed for academics or entrepreneurs from any discipline who teach or have a desire to teach entrepreneurship. While in the program, you will learn alongside like-minded educators dedicated to exploring innovative, bold, and impactful ways of shaping the next generation of entrepreneurial leaders (http://www.babson.edu/executive-education/education-educators/Pages/home.aspx).
- ***Lean Launchpad® Educators Seminar***: The Lean LaunchPad® Educators Seminar is a 2.5-day program designed for entrepreneurship faculty who want to learn the basics of integrating Lean Startup principles into their teaching. You will learn key techniques for building a curriculum that you can adapt to your institutional environment and needs and put to immediate use (https://venturewell.org/lean-launchpad/).
- ***United States Association of Small Business and Entrepreneurship (USASBE) Annual Conference***: In addition to engaging in entrepreneurship research, this conference offers multiple professional development resources pertaining to education. USASBE also co-sponsors the Go Deep Entrepreneurship Teaching & Learning Scholars Program (http://www.usasbe.org/).
- ***VentureWell Open Conference***: Formerly known as the National Collegiate Inventors and Innovators Alliance, VentureWell Open brings together students as well as entrepreneurship educators to "share stories, start new collaborations and learn best practices. (https://venturewell.org/about-open/).
- ***Deshpande Symposium for Innovation & Entrepreneurship in Higher Education***: The Deshpande Symposium is a gathering of like-minded practitioners focused on accelerating innovation and entrepreneurship across the college and university environment. The Symposium provides a venue for attendees to learn from each others' experiences; to gain wider exposure for their ideas, successes and start-ups; and to collaborate on activities that increase innovation and entrepreneurship in their academic communities (http://www.deshpandesymposium.org/).
- ***National Association for Community College Entrepreneurship (NACCE) Annual Conference***: The NAACE annual conference started in 2003 and provides an opportunity for community college stakeholders to network and obtain professional development towards incorporating the entrepreneurial method in the community college classroom (http://www.nacce.com).
- ***Intrapreneurship Conference***: Founded in 2011, this organization offers several global opportunities for corporate innovators to obtain practical insights into trends, best practices, and networking with other successful entrepreneurial leaders. The target audience includes innovations across companies and industries (https://www.intrapreneurshipconference.com).

7.4 Organizations, Initiatives, and Movements

Finally, what are the organizations, initiatives, and movements that you want to keep a pulse on, as you strive to nurture the entrepreneurial mindset of your students? Which should you consider membership in and also follow? Here, we offer our list of recommendations—again, many from the engineering arena, but also several complimentary organizations that are sure to be insightful.

- Kern Entrepreneurial Engineering Network (KEEN) (http://engineeringunleashed.com/keen/)
- National Center for Engineering Pathways to Innovation (Epicenter) (http://epicenter.stanford.edu/)
- Worldwide CDIO (Conceive, Design, Implement, Operate) Initiative (www.cdio.org)
- American Society for Engineering Education (ASEE) - Entrepreneurship and Innovation Division (https://sites.asee.org/ent/)
- Stanford d.School (https://dschool.stanford.edu/)
- NSF I-Corps (https://www.nsf.gov/i-corps)
- Small Business Innovation Research (SBIR) and Small Technology Transfer Research (STTR) Grants (https://www.sbir.gov/about)
- Midwest Entrepreneurship Engineering Network (http://tec.illinois.edu/about/national-collaborations/meen.asp)
- National Academy of Engineering (NAE) Grand Challenges for Engineering (http://www.engineeringchallenges.org/)
- Venturewell (https://venturewell.org/)
- Kauffman Foundation (http://www.kauffman.org/)
- Stanford eCorner (https://ecorner.stanford.edu/)
- Network for Teaching Entrepreneurship (http://www.nfte.com)
- World Series of Innovation (http://innovation.nfte.com/).

Chapter 8
Resistance and Change: Making A Case for the Entrepreneurial Mindset

Author Reflection

If you have made it this far into the book, it is likely you don't need much more convincing to incorporate the entrepreneurial mindset into your engineering classroom. However, it's very likely that your faculty peers and even administration may not yet have caught onto the great benefits and need to incorporate the entrepreneurial mindset into engineering education. Put blankly—they haven't yet "drank the kool-aid" and may not even know where to start. If you want to better prepare yourself for driving change in the potential face of resistance, this chapter is for you.

The purpose of this chapter is to identify the challenges, barriers, and how to make a case for the entrepreneurial mindset.

8.1 Challenges and Barriers

When making the case for the entrepreneurial mindset at your institution, there are two realities to bear in mind. First, change happens, but takes time. We must be patient, but also persevere and make constant strides and steps forward. And, second, that change and resistance typically seem to go hand-in-hand, in that most people are resistance to change to some extent. Thus, it is important to question and understand why there is resistance.

In her Harvard Business Review article, Kantar (2012) identifies the top ten reasons why people resist change: loss of control, excess uncertainty, surprise, everything seems different, loss of face, concerns regarding competence, more work, ripple effect, past resentments and sometimes the threat they are anticipating is real. She further posits that identifying the particular source of resistance is key to addressing it. It usually has to do with (a) being comfortable with where they are and (b) the comfortableness and effort it takes to learn and get used to something

L. Bosman and S. Fernhaber, *Teaching the Entrepreneurial Mindset to Engineers*, DOI 10.1007/978-3-319-61412-0_8

new. Specific to the entrepreneurial mindset, some faculty may feel out of their element in that they may be lacking in entrepreneurship experience and do not really understand it (Zappe et al. 2013). For others, entrepreneurship is seen as important, but just lower on the priority list of learning objectives where curriculum tends to be content dense as it is (Peterfreund 2013; Weilerstein and Byers 2016). Whatever the cause for resistance may be, we must be able to recognize the source of the resistance and address it proactively if we want to move forward both efficiently and effectively. Knowing this, you can help make the case for integrating an entrepreneurial engineering perspective at your institution *while* making it easier for them. Fortunately, there is a lot of data to back this up—which helps makes the individual resistance to change to understand that your arguments are both valid and needed. In essence, you are helping them do their homework.

What does that mean for you, if you are a champion for entrepreneurial engineering at your institution? Fortunately, university leadership appears to be supportive of entrepreneurial education for engineering students, and serve as champions (Byers et al. 2013). This appears to be an opportunity to leverage this support. Additionally, research suggest that best practices and professional development opportunities need to compliment awareness to obtain faculty buy-in. Borrego et al. (2010) found that the highest curriculum change adoption rates were found when the job could be completed by individuals or small teams with limited requirement for departmental coordination.

What are the challenges faced by engineering faculty in integrating entrepreneurship programming? A survey by Zappe et al. (2013) identify the top four challenges to be (1) institution's bureaucracy, (2) students' prior knowledge, (3) different type of workload involved with teaching entrepreneurship, and (4) budget constraints.

There are many stakeholders involved when considering incorporating the entrepreneurial mindset into engineering curriculum, such as students, faculty, university, industry, and society. When making the case for the entrepreneurial mindset, one must consider the motivations and perspectives of the individual faculty member and their capacity for making academic change happen. There are many benefits for faculty who incorporate the entrepreneurial mindset into the engineering classroom.

First, engineering educators have the potential to make a difference in the life of a student and economic impact to society when a student actual stumbles upon a validated design innovation. Second, engineering educators benefit because of the increase in student motivation for learning the subject, allowing students to see the value and understand the practical purpose of the topic and how it contributes to the economy, and increasing teacher evaluations. Third, engineering educators that fail to get on board will likely get left behind; the revolution in engineering education is coming and engineering educators need to be prepared to explore and promote new ways of teaching and learning. Fourth, entrepreneurially minded engineering students make great student researchers and student research assistants, especially for NSF I-Corps and Small Business Technology Transfer (STTR) grant projects. Finally, engineering educators that teach the entrepreneurial mindset will naturally increase their own entrepreneurial mindset, ultimately optimizing the work–life balance to fit their goals.

8.2 Making Your Case

While we hope this book is helpful in equipping you, as the engineering educator to develop the entrepreneurial mindset in your students, we also recognize the need to be able to "get others on board!" Thus, we end with what we feel are the top arguments for convincing faculty peers to integrate the entrepreneurial mindset into the engineering curriculum, persuading students to embrace the entrepreneurial mindset, and encouraging administration to dedicate resources towards building an ecosystem promoting the entrepreneurial mindset in students, faculty, and staff. We hope that you can use these to help make your case, and add on to them as you develop your entrepreneurial mindset.

How to Convince Faculty Peers to Incorporate the Entrepreneurial Mindset into Curriculum?

1. ***More Engaged Students***: Incorporating the entrepreneurial mindset not only makes sense, but can help motivate your student to be more motived and engaged in learning, and result in higher teaching evaluations (Boice 2000). Integrating the entrepreneurial mindset into the engineering classroom allows students to see the value and understand the practical purpose of the topic and how it contributes to the economy. As a result, students are less like to skip class or sleep in class, and are likely to show an increased interest and understanding in the topic as demonstrated through teacher evaluations.
2. ***Funding Opportunities***: Entrepreneurship drives economic development, which is why it is supported by federal, state, and philanthropic organizations. Engineering faculty who incorporate the entrepreneurial mindset into learning opportunities (e.g., curricular, non-curricular, formal, informal) and turn those learning opportunities into commercializable products or services are meeting the goals set forth by a variety of funding agencies including VentureWell, KEEN, Lemelson Foundation, Kauffman Foundation, NSF I-CORPS (and NSF I-CORPS for Learning), STTR, and Small Business Innovation Research (SBIR).
3. ***Research Opportunities***: Engineering faculty who incorporate the entrepreneurial mindset into the classroom and assess the learning outcomes have the opportunity to publish the research findings in journals such as Journal of Engineering Entrepreneurship, American Society for Engineering Education (ASEE) Conference Proceedings on Entrepreneurship and Engineering Innovation, Journal of Engineering Education, Advances in Engineering Education, Journal of Entrepreneurship Education, and Experiential Entrepreneurship Exercises.
4. ***Networking***: Engineering faculty interested in incorporating the entrepreneurial mindset into their curriculum have immediate access to a community of

like-minded individuals and organizations. Faculty can get involved by participating in a number of nationally recognized entrepreneurially minded workshops and conferences including Venturewell's Open Conference, the KEEN National Conference, and the ASEE Conference division on Entrepreneurship and Engineering Innovation. Also, many college campuses and/or cities offer a local startup community including access to programs such as pitch events, business model competitions, and accelerator startup workshops.

5. ***Optimize Work–Life Balance***: Engineering faculty that teaches the entrepreneurial mindset will naturally increase his or her own entrepreneurial mindset. Entrepreneurially minded faculty are capable of recognizing and capitalizing on opportunity in all facets of their lives, from personal lives to research lives to teaching lives to career lives because there are "customers" all around us. Entrepreneurially minded faculty excels at optimizing the work–life balance to fit the goals set for themselves.

How to Persuade Students to Embrace the Entrepreneurial Mindset

1. ***Increases Value to Employers***: Engineering students need entrepreneurial skills to help address pressing needs and employer demands in the workplace (Byers et al. 2013). Taking courses integrating the entrepreneurial mindset can help boost job prospects, as these are the skills that employers want (Duval-Couetil and Wheadon 2013).
2. ***Learn How to Start a Company***: Enrolling in courses incorporating the entrepreneurial mindset can provide students with just the amount of exposure and experience to start their own business, perhaps even while still working on their degree. Here, they have access to mentors and resources, which can sometimes be difficult to obtain after they graduate and leave the university setting.
3. ***Try It Before You Buy It***: Obtaining an engineering degree can be quite demanding because students are taking a full load of rigorous quantitative focused courses. Taking engineering electives incorporating the entrepreneurial mindset will give students an opportunity to "test-out" an interest in starting their own company prior to investing time and money into an Entrepreneurship and Innovation Minor.
4. ***Creative Outlet***: A bachelor's degree in engineering typically requires taking quite a few calculus, physics, and chemistry courses, in addition to the core engineering courses. These quantitative focused courses can be mentally taxing as students can spend hours on end working on trying to get the "right answer" for one homework problem. Incorporating the entrepreneurial mindset can offer students a glimpse into the relevancy of a course and also a creative outlet that compliments the technical skillset.
5. ***Optimize Student-Life Balance***: Students that develop the entrepreneurial mindset are capable of recognizing and capitalizing on opportunity in all facets of their lives, from personal lives to student lives to career lives because there are "customers" all around us. Entrepreneurially minded students excel at optimizing the student-life balance to fit the goals set for themselves.

How to Encourage Administration to Dedicate Resources Towards Building an Ecosystem Promoting the Entrepreneurial Mindset

1. ***Increase in Retention Rates***: Integrating entrepreneurship into engineering can boost retention rates & GPAs (Ohland et al. 2004). With enrollment into engineering programs increasing worldwide, there is a need to engage and increase retention—which we argue can be done via entrepreneurial mindset as it helps emphasize the real world applicability of engineering and better prepares our students for what they already see and are exposed to in the 'real world.'
2. ***Making Your Mark***: Olin College of Engineering and the iFoundry at the University of Illinois are well known for how they do engineering education differently (Goldberg and Somerville 2014). Transforming education through the entrepreneurial mindset is on trend nationally and many institutions are moving toward the entrepreneurial ecosystem model. Most every new university administrator wants to "make their mark" on the organization and the entrepreneurial mindset could be just the thing to propel an administrator's career forward.
3. ***Student Placement Rate and University Rankings***: Recent research (Nair et al 2009) suggests that industry employers are looking for students that possess both the "engineering fundamentals" and professional skills, such as communication, teamwork, and critical thinking. Integrating the entrepreneurial mindset into the student learning experience offers an easy way to meet the ever growing demands and desires of industry, ultimately increasing student placement rate.
4. ***Increase Internal and External Collaborators***: A university's strategic plan (and following through action) can provide a starting point for building an entrepreneurial ecosystem within the campus arena and throughout the community. Entrepreneurial initiatives commonly require and encourage participation from across colleges within a university setting. Furthermore, the support of an entrepreneurial ecosystem has the potential to draw attention from the local startup community, innovation focused non-profits, philanthropic organizations, and large employers looking to invest in entrepreneurial activities.
5. ***Potential New Revenue Stream***: If a student or faculty creates a company using university resources, there is the potential for the university to own part of the Intellectual Property (IP) and benefit from a financial perspective. Thus, training engineering faculty and students in the entrepreneurial mindset should result in increased patent development, grant funding, industry-sponsored faculty research projects, and commercialized products and services. In addition, minors in Entrepreneurship and Innovation provide a supplementary revenue stream sought after by many universities.

References

Boice, R. (2000). *Advice for new faculty members*. Nihil nimus, Allyn & Bacon.

Borrego, M., T.S. Hall, and J.E. Froyd. 2010. Diffusion of engineering education innovations: A survey of awareness and adoption rates in US engineering departments. *Journal of Engineering Education* 99 (3): 185–207.

Byers, T., T. Seelig, S. Sheppard, and P. Weilerstein. 2013. Entrepreneurship: Its role in engineering education. *The Bridge on Undergraduate Engineering Education* 43 (2): 35–40.
Duval-Couetil, N., and J. Wheadon. 2013. The value of entrepreneurship to recent engineering graduates: A qualitative perspective. Frontiers in Education Conference, 2013 IEEE, IEEE.
Goldberg, D.E., and M. Somerville. 2014. *A whole new engineer*. USA: Three Joy Associates Inc.
Kantar, R.M. 2012. Ten reasons people resist change. *Harvard Business Review.*
Nair, C.S., et al. 2009. Re-engineering graduate skills—a case study. *European Journal of Engineering Education* 34 (2): 131–139.
Ohland, M.W., et al. 2004. The effect of an entrepreneurship program on GPA and retention. *Journal of Engineering Education* 93 (4): 293–301.
Peterfreund, A.R. 2013. Epicenter baseline survey: Report of Findings. Retrieved from: http://sagefoxgroup.com/epicenter.
Weilerstein, P. and Byers, T. 2016. Guest editorial: Entrepreneurship and innovation in engineering.
Zappe, S., K. Hochsedt, E. Kisenwether, and A. Shartrand. 2013. Teaching to innovate: Beliefs and perceptions of instructors who teach entrepreneurship to engineering students. *International Journal of Engineering Education* 29 (1): 45–62.

Part IV
Example Engineering Curriculum Incorporating the Entrepreneurial Mindset

Chapter 9
Example Engineering Curriculum: Online Discussion Prompts

Author Reflection

Several years ago, I attended a conference and found myself chatting with fellow engineering faculty during an after-hours event. Somehow, the topic turned to our first time teaching at the college level. One of the young professor's story went something like this: "As a graduate student, I was required to teach an engineering statistics course. I had no idea what I was doing and had no idea who to ask for help. I decided to stick with the current book, and after much Googling, I was able to find a previous syllabus using the same book, so I went with it. I was lucky enough to find the publisher's website, where I discovered that the PowerPoint lectures and test-banks were already created. Score! At this point I thought to myself that teaching can't be that difficult. As it turns out, the lectures bored my students to death and half the time they didn't show up so they ended up performing poorly on the exams".

If you have a Ph.D. in Engineering AND during your doctoral studies you received training in how to teach at the college level, you are a rarity. Many faculty go through the doctoral program with a focus on how to be a 'good researcher' not necessarily how to be a 'good teacher'.

The purpose of this chapter is to provide examples on how to incorporate the entrepreneurial mindset into existing, engineering curriculum using online discussion prompts.

L. Bosman and S. Fernhaber, *Teaching the Entrepreneurial Mindset to Engineers*, DOI 10.1007/978-3-319-61412-0_9

Example 1. Assigned Readings and Standard Prompts: Opportunity Discovery and the Utility Grid

Author: Dr. Nathan Weise
Course: Renewable and Legacy Electric Energy Systems Analysis
Learning Activity Approach: Assigned Readings and Standard Prompts

Description

This course is an elective, which explores elements of electric power systems with a specific focus on system design, analysis, and management. This course used a variety of articles and writing prompts to promote and encourage an understanding of issues, challenges, and key drivers facing the utility grid.

Learning Objectives

1. Interpret knowledge about the current status and new developments in the field of electric energy systems.
2. Critique and explore a contrarian view of positions and perspectives in the field of electric energy systems.

Learning Activities

Students had 1 week to read the article and respond to the writing prompts using the online learning management system. Students were given 2 points for an excellent response and 0 points for no response. A total of 8 readings and writing prompts were given throughout the 16-week academic semester. The writing prompts for all readings were as follows:

- What are you curious about regarding the grid and why?
- What did you find interesting regarding the history of the grid?
- What did you find interesting about planning and natural disasters?
- Discuss the challenges facing the grid. What are the keys to becoming the utility of the future?

Week 1: *The Utility and Grid of the Future*

- Aguero, J. R., et al. (2016). "The Utility and Grid of the Future: Challenges, Needs, and Trends." IEEE Power and Energy Magazine 14(5): 29–37.

Week 2: *Becoming the Utility of the Future*

- Brown, R. E., et al. (2016). "Becoming the Utility of the Future: Risks and Opportunities." IEEE Power and Energy Magazine 14(5): 57–65.

Week 3: *An Introduction to Electrical Diagnostic Testing of Power Transformers*

- Dupuis, B. (2016). "An introduction to electrical diagnostic testing of power transformers." IEEE Power and Energy Magazine 14(5 Supplement): 38–45.

Week 4: The China Southern Power Grid—Solutions to Operations Risk and Planning Challenges

- Zhou, H., et al. (2016). "The China Southern Power Grid: Solutions to Operation Risks and Planning Challenges." IEEE Power and Energy Magazine 14(4): 72–78.

Week 5: Frequency Disturbances During the Super Bowl

- Allen, E. H., et al. (2016). "Frequency Disturbances During the Super Bowl: It's More Than Just What's on the Field." IEEE Power and Energy Magazine 14(6): 52–58.

Week 6: Tales of Power System Failures

- Waldele, R. W. (2016). "Tales of Power System Failures: A Look at the Unusual, Strange, and Downright Bizarre Causes." IEEE Power and Energy Magazine 14(6): 18–23.

Week 7: A Colorful Blackout

- Guillon, S., et al. (2016). "A Colorful Blackout: The Havoc Caused by Auroral Electrojet Generated Magnetic Field Variations in 1989." IEEE Power and Energy Magazine 14(6): 59–71.

Week 8: Watch Out for Flooding

- Allen, E. H. (2016). "Watch Out for Flooding: When the Power System Created a Weather Disaster." IEEE Power and Energy Magazine 14(6): 35–39.

Example 2. Discussion Prompts to Connect Course Topics to Real-World Applications: Engineering Discovery and Innovation Exploration

Authors: Dr. Lisa Bosman and Dr. Hyunjae Park
Course: Introduction to Engineering
Learning Activity Approach: Discussion Prompts to Connect Course Topics to Real-World Applications

Description

This course is a requirement for all engineering majors. Some of the objectives include providing an overview to a variety of engineering disciplines, exploration of the many roles engineers play in society, application of professional skills, and lays the foundation for innovation and the engineering design process. This course used a variety of discussion prompts to connect real-world applications to the course objectives.

Learning Objectives

1. Describe the varying roles of engineering disciplines within society.
2. Critique and explore new developments in the field of engineering.

Learning Activities

Each week of the 16-week semester, with the exception of midterm and final exam weeks, students were required to reply to a variety of discussion prompts. All discussion prompts required an initial post and a response to a peer's post, which needed to be completed within a week.

Week 1: *Introduction*

Part 1: Upload a profile picture of yourself.
Part 2: Create a post answering these questions.

- Why do you want to be an engineer and/or study engineering?
- What type/kind of engineer do you want to be (in the future)?
- What's your goal as an (future) engineer?

Part 3: Respond to a peer's post welcoming them to the class.

Week 2: *Historical Engineers*

Part 1: Many of the greatest advancements in history have come about as the direct result of those working as engineers. Engineers provide us with practical solutions for a host of problems, as well as advance practical science and technology. They take theories and ideas, and often turn them into working principles and products that better our lives. From the compound pulley system invented by the great Greek engineer Archimedes, to the tall buildings and air-conditioned comfort we enjoy today, engineers have been at the forefront of our technological advancement. Identify whom you believe to be one of the greatest engineers and state why.

Part 2: Select two great engineers posted in the discussion. Assuming they were alive at the same time and capable of collaboration, what could they have accomplished?

Week 3: Math Importance

Part 1: Calculus is the study of how things change. It provides a framework for modeling systems in which there is change, and a way to deduce the predictions of such models. Provide an example of where an engineer might use calculus in a real-world application and explain how it is used.
Part 2: Select a peer's post. Respond with a prediction of how calculus might be used in this real-world application 20 years from now.

Week 4: Nuclear Engineering

Part 1: Nuclear engineers research and develop the processes, instruments, and systems used to derive benefits from nuclear energy and radiation. Many of these engineers find industrial and medical uses for radioactive materials—for example, in equipment used in medical diagnosis and treatment. Nuclear engineers typically work in offices; however, their work setting varies with the industry in which they are employed. Most nuclear engineers work full time. Search the Internet to find a nuclear engineering focused design you think is especially innovative. Explain the product or service and how it adds value to the customer.
Part 2: Select a peer's post and comment on how current market trends might impact the innovation in the future.

Week 5: Biomedical Engineering

Part 1: Biomedical engineers combine engineering principles with medical and biological sciences to design and create equipment, devices, computer systems, and software used in health care. Search the Internet to find a biomedical engineering focused design you think is especially innovative. Explain the product or service and how it adds value to the customer.
Part 2: Select a peer's post and comment on how current market trends might impact the innovation in the future.

Week 6: Industrial Engineering

Part 1: Industrial engineers find ways to eliminate wastefulness in production processes. They devise efficient systems that integrate workers, machines, materials, information, and energy to make a product or provide a service. Depending on their tasks, industrial engineers work either in offices or in the settings they are trying to improve. For example, when observing problems, they may watch workers assembling parts in a factory. When solving problems, they may be in an office at a computer, looking at data that they or others have collected. Search the Internet to find an industrial engineering focused design you think is especially innovative. Explain the product or service and how it adds value to the customer.

Part 2: Select a peer's post and comment on how current market trends might impact the innovation in the future.

Week 7: Civil Engineering

Part 1: Civil engineers design, build, supervise, operate, and maintain construction projects and systems in the public and private sector, including roads, buildings, airports, tunnels, dams, bridges, and systems for water supply and sewage treatment. Search the Internet to find a commercially available civil engineering focused design you think is especially innovative. Explain the product or service and how it adds value to the customer.
Part 2: Select a peer's post and comment on how current market trends might impact the innovation in the future.

Week 8: Midterm Week (No Discussion Prompt)

Week 9: Chemical Engineering

Part 1: Chemical engineers apply the principles of chemistry, biology, physics, and math to solve problems that involve the production or use of chemicals, fuel, drugs, food, and many other products. They design processes and equipment for large-scale manufacturing, plan and test production methods and byproducts treatment, and direct facility operations. Chemical engineers work mostly in offices or laboratories. They may spend time at industrial plants, refineries, and other locations, where they monitor or direct operations or solve onsite problems. Nearly all chemical engineers work full time. Search the Internet to find a chemical engineering focused design you think is especially innovative. Explain the product or service and how it adds value to the customer.
Part 2: Select a peer's post and comment on how current market trends might impact the innovation in the future.

Week 10: Mechanical Engineering

Part 1: Mechanical engineering is one of the broadest engineering disciplines. Mechanical engineers design, develop, build, and test mechanical and thermal sensors and devices, including tools, engines, and machines. Search the Internet to find a mechanical engineering focused design you think is especially innovative.

Explain the product or service and how it adds value to the customer.

Part 2: Select a peer's post and comment on how current market trends might impact the innovation in the future.

Week 11: Agricultural Engineering

Part 1: Agricultural engineers attempt to solve agricultural problems concerning power supplies, the efficiency of machinery, the use of structures and facilities, pollution and environmental issues, and the storage and processing of agricultural products. Agricultural engineers work in a variety of industries. Some work for the federal government, and others provide engineering contracting or consultation services, or work for agricultural machinery manufacturers. Although they work

mostly in offices, they also may spend time traveling to agricultural settings. Search the Internet to find an agricultural engineering focused design you think is especially innovative. Explain the product or service and how it adds value to the customer.
Part 2: Select a peer's post and comment on how current market trends might impact the innovation in the future.

Week 12: *Electrical Engineering*

Part 1: Electrical engineers design, develop, test, and supervise the manufacturing of electrical equipment, such as electric motors, radar and navigation systems, communications systems, and power generation equipment. Search the Internet to find an electrical engineering focused design you think is especially innovative. Explain the product or service and how it adds value to the customer.
Part 2: Select a peer's post and comment on how current market trends might impact the innovation in the future.

Week 13: *Computer Engineering*

Part 1: Computer hardware engineers research, design, develop, and test computer systems and components such as processors, circuit boards, memory devices, networks, and routers. These engineers discover new directions in computer hardware, which generate rapid advances in computer technology. Computer hardware engineers usually work in research laboratories that build and test various types of computer models. Most work in high-tech manufacturing firms. Search the Internet to find an computer engineering focused design you think is especially innovative. Explain the product or service and how it adds value to the customer.
Part 2: Select a peer's post and comment on how current market trends might impact the innovation in the future.

Week 14: *Engineering Impact*

Part 1: Team building is important in any environment which requires the coordination and integration of multiple activities. It is especially crucial in a technical engineering environment where projects are often highly complex and require the integration of many functional specialties. For these projects, it is necessary for the managers and their lead engineering personnel to cross organizational boundaries and work with resources and personnel over whom they have little or no formal authority. Such a team must have a capacity for innovatively transforming a set of technical objectives and requirements into specific products, system concepts, or services that compete favorably against other available alternatives. Take into consideration your role in the final project and identify how your role contributed to the team's success.
Part 2: Select a peer's post where you have played a similar role but in a different setting (e.g., job, family, club, sports, extracurricular activities, etc.…) and describe your experience.

Week 15: *What Is Engineering?*

Part 1: Unfortunately, engineers haven't done a good job in getting the word out about what they really do (perhaps because they're too busy doing it!). As a result, a lot of misconceptions exist about what engineering is and what engineers do, with people believing engineers to be anything from locomotive train operators to mechanics/technicians, to construction supervisors, to NASA personnel, to generic "computer people". How would you explain the role of engineers to an adult? To a child?
Part 2: Select a peer's post. Assume you are that unknowing adult or child. What types of follow-up questions would you ask to learn more about engineering?

Week 16: *Finals Week (No Discussion Prompt)*

Example 3. Discussion Prompts Requiring References: Biocomputing and Medical Innovation Exploration

Author: Dr. Robert Scheidt
Course: Biocomputers Design Lab 1
Learning Activity Approach: Discussion Prompts Requiring References

Description

This course is a requirement for all Biocomputing Biomedical Engineering majors and focuses on medical devices utilizing embedded hardware and software. Since the course offers hands-on practice with software design and validation, the online discussion prompts offers a complimentary extension of the lab environment into current real-world Biocomputing applications.

Learning Objectives

1. Describe the role of technology in medical innovations.
2. Critique and explore the development of technology-based medical devices.

Learning Activities

This activity required students to reflect upon, write a response, and respond to peer posts categorized into 5 major themes. Students were assigned two points for each well-thought-out response to a prompt and for each well-considered comment to a peer's response. One point was given if the effort was weak. No points for missing or late efforts. This counted for 10% of the course grade.

Prompt 1: *Agile Software Development*

Discussions in this course provide opportunities to analyze larger issues related to the development of microcontroller-based medical devices. Be sure to support your comments with references to the module resources, as well as to relevant examples and/or experiences you have had. Start a new thread to submit your initial response to the prompt. Then, compose two responses to classmate comments by Wednesday. For this discussion, respond to the following prompts:

Agile software development describes a set of principles that foster iterative, incremental and evolutionary development, continuous improvement and rapid, flexible response to change. Agile development favors coding over documentation. Agile welcomes changing requirements, even late in development. By contrast, medical device developers that seek to obtain the FDA approval to market their wares (and to survive an FDA audit) are required to have extremely well-documented design history files and detailed documentation regarding testing activities that establish the safety and efficacy of their products. On the face of it, Agile development appears incompatible with medical device development. Respond to the following:

- In your opinion, are Agile methods compatible with medical device development? Why or why not?
- What is the FDA's perspective on the application of Agile methods to the development of embedded software?
- Identify one medical device developer that claims that they employ Agile methods. How do they manage to do so?

Prompt 2: *Biomedical Engineering Society Code of Ethics*

Discussions in this course provide opportunities to analyze larger issues related to the development of microcontroller-based medical devices. Be sure to support your comments with references to the module resources, as well as to relevant examples and/or experiences you have had. Start a new thread to submit your initial response to the prompt. Then, compose two responses to classmate comments by Wednesday.

For this discussion, respond to both of the following:

Part 1: Analyze the four cannon of the Biomedical Engineering Society Code of Ethics. Identify the following:

1. The three enumerated elements you think are the most important.
2. The one or two you think might be the most challenging for work-a-day engineers.
3. The one or two you think might be the most challenging for engineering firm owners or upper level engineering managers.

Provide a brief explanation of the selections you made for each of the items above.

Part 2: Select a biomedical company from the Fortune 500 list. Examine the website of that company you select. Describe the company's corporate responsibility policy and how it relates (if at all) to the rights of patients and the larger society in regard to cost, availability, and delivery of health care. (If the company does not have a corporate responsibility policy, suggest one for it that includes elements related to the BMES code of ethics.) Is the policy proactive or reactive? Is the company actively engaged in corporate citizenship in a meaningful way or is it only giving lip service? Make the case for why the company you choose is a responsible corporate citizen (or not).

Prompt 3: *Engineering Ethics and Therac-25*

Discussions in this course provide opportunities to analyze larger issues related to the development of microcontroller-based medical devices. Be sure to support your comments with references to the module resources, as well as to relevant examples and/or experiences you have had. Start a new thread to submit your initial response to the prompt. Then, compose ONE response to classmate comments by Wednesday.

For this discussion, respond to the following:

The Therac-25, a radiation therapy machine produced by Atomic Energy of Canada Limited (AECL) in 1982, was involved in at least six accidents between 1985 and 1987, in which patients were given massive overdoses of radiation. Because of concurrent programming errors, the Therac-25 sometimes gave its patients radiation doses that were hundreds of times greater than normal, resulting in death or serious injury. The problem was difficult to reproduce (it was a classic "Heisenbug"). The Therac-25 accidents highlighted the dangers of software control of safety-critical systems, and have become a standard case study in medical device development. Review the 1993 paper by Leveson and Turner and then respond to the following prompts.

1. Leveson describes several software "bugs" resulting from "concurrent programming errors". Choose one at the heart of the Therac-25 problem and describe the failure mode that it caused.
2. To what extent did software documentation (or a lack thereof) contribute to the Therac25 problems?
3. When AECL was notified of the problem you chose, describe the steps it took to mitigate the problem. Was this response satisfactory? If so, defend your claim. If not, describe steps AECL could have taken to improve its response.
4. Consider you are a newly graduated engineer hired by AECL to resolve their real-time problems. Describe how you might use resources provided by uC/OS-II to mitigate the bug you described in part 1.

In your response to the four prompts, cite any literature and/or websites you used to support your stance.

Prompt 4: *Innovations in Expansion Modules*

Discussions in this course provide opportunities to analyze larger issues related to the development of microcontroller-based medical devices. This week, the focus of discussion is on novel devices that can expand the inherent capabilities of a standard microcontroller using SPI or IIC interfaces. Be sure to support your comments with references to relevant web sources and/or experiences you have had. Start a new thread to submit your initial response to the prompt by end of day, Wednesday. Then, compose TWO responses to classmate comments by the following Wednesday.

For this discussion, respond to the following:

In recent years, there has been an explosion of "expansion modules" available to the embedded systems designer to enhance the inherent capabilities of their designs. Examples include ultra high-resolution 24–32-bit A/D convertors (http://www.ti.com/lit/ds/symlink/ads1281.pdf) and inexpensive wifi modules (https://www.sparkfun.com/products/13678). Go online and do a web search to identify two or three SPI or IIC expansion chips or modules that you consider to be cool or exciting, then respond to the following prompt.

1. Provide a web link to each of your selected devices.
2. Describe what each of your selected devices do, and why you think it is exciting.
3. Describe a potential embedded application that you could integrate each of your selected devices into.

In your response to the three prompts, cite any additional literature and/or websites you used to support your ideas.

Prompt 5: *Understanding Your Archetypical Customer*

Discussions in this course provide opportunities to analyze larger issues related to the development of microcontroller-based medical devices. For this last prompt, the focus of discussion is on the customer discovery process—what some call "Empathy" as it relates to "Design Thinking". Specifically, "get out of the building" and interview an archetypical customer to test a specific hypothesis about your project (the AutoGoni). Try to obtain a deep insight into user behavior rather than user opinion. Be sure to report out key takeaways from the interview, highlighting both what worked and what did not. Start a new thread to submit your summary of your customer/user interview by end of day, Wednesday. Then, in the following week, compose TWO responses to classmate comments.
For this discussion, respond to the following:

In class, we viewed several videos from Steve Blank describing how to plan and execute successful "customer discovery" interviews. After formulating in your mind a picture of the "archetypical customer", identify a specific hypothesis regarding what pressing need your AutoGoni will satisfy for your customer, and then "Get out of the building!" to test that hypothesis.

1. Describe the specific hypothesis you set out to test in your interview.
2. Describe your interviewee and how this individual fits with your view of the archetypical customer/user.
3. Describe the key takeaways from the interview. What did you learn related to your hypothesis? What did you learn tangential from your hypothesis?
4. During the interview process, what worked for you? What did not work?

In your response to the four prompts, try to be as detailed as possible so that your classmates can learn from your experiences.

Example 4. Speaker-based Discussion Prompts: Real-World Applications of Water Resources Challenges

Author: Dr. Brooke Mayer
Course: Environmental Engineering Seminar
Learning Activity Approach: Speaker-Based Discussion Prompts

Description

This course is a noncredit seminar requirement for all Environmental Engineering majors. Topics include air pollution, wastewater treatment, urban hydrology, storm water management, and hazardous waste management. The incorporation of online discussions offered a unique opportunity for students to dig deeper to learn more about real-world water resource challenges.

Learning Objectives

1. Describe the topic areas of interest related to environmental engineering and water resources.
2. Critique and explore the challenges around water resources, treatment, and pollution.

Learning Activities

Online Discussion: Each week, students post an initial post and a response based on the discussion prompt. Example discussion prompts, many of which incorporate in-class speakers, are provided.

Prompt 1: *Decentralized Versus Centralized Treatment*

In the U.S., the most widely used/accepted approach to treatment of domestic wastewater is centralized treatment in the form of facilities designed to use a combination of physical, chemical, and biological treatments to mitigate contaminants in the wastewater collected from a number of households. However, more than 1 in 5 households in the U.S. are instead served by septic systems. This is an example of decentralized treatment, where instead of collecting all of the wastewater in a sewer and treating it in one centralized location, the wastewater is treated in smaller volumes at different locations near where it is generated. In many developed countries around the world where there is no existing sewage treatment, there is great interest in establishing decentralized versus centralized wastewater treatment (for example, using single family unit UASBs, like the speaker talked about). If you were charged with designing the wastewater treatment system for a brand new city in the U.S., would you opt for centralized or decentralized treatment? Defend your answer. Would your response change if you were located in Brazil, for example?

Prompt 2: *Energy Efficiency in Wastewater Treatment*

The speaker described a number of different options for improving energy efficiency at wastewater treatment plants. If you had the opportunity to pick one (and only one) energy efficiency updated to an existing wastewater treatment plant, what would it be and why? If many different energy efficiency options were implanted in cohesive strategy, do you think it's feasible for wastewater treatment plans to be energy positive (meaning they produce more energy than they consumer, and put some back into the grid) rather than energy negative (they use more energy than produce)?

Prompt 3: *Nanomaterials*

Interest in applications using nanoparticles has greatly increased in recent years. Specifically, use of nanomaterials in water treatment settings has also garnered interest. For example, the research that the speaker discussed shows greatly improved efficiency in degrading highly oxygenated contaminants (e.g., bromate) in water using nanomaterials. To date, such materials have not been used in real-world water treatment, with one potentially contributing factor being that the human health implications of ingestion of nanoparticles are not yet well defined. Do you think that use of possibly powerful nanotechnology is the future of water/wastewater treatment or do you believe it will not come to fruition (or somewhere in between)? Explain and defend your answer considering alternate viewpoints.

Prompt 4: *Why Is Ecological Diversity Important?*

The speaker discussed the mechanisms of maintaining ecological diversity in tropical forests, with an emphasis on trees and vines. Describe a particular setting wherein ecological diversity is important in relation to environmental engineering goals. What makes it so important in your example? What drives the diversity? Are there ways to ensure that diversity is maintained? In your initial post, be sure to use an example that has not already been used by a classmate (check the existing topic postings first). In your response, compare and contrast the example you identified with another person's.

Prompt 5: *Waste Versus Resource*

The paradigm of wastewater treatment is shifting from a focus solely on waste mitigation toward recovery of valuable resources. In fact, organizations such as WEF are now calling wastewater treatment facilities "Water Resource Recovery Facilities (WRRFs)" rather than wastewater treatment plants (WWTPs). Recoverable products could include water, energy, nutrients, metals, etc.... (see nice ES&T article on this topic). Do you think that recover of any/all of these resources is possible in the future? Do you think there would be a sufficient market for such products such that WRRFs could actually make money? How do you think the public would view this process (i.e., is there a "yuck factor")? What are some constraints or incentives to either discourage or encourage increased efforts at recovery of resources from wastes?

Prompt 6: *Little Baby Digesters*

In wastewater treatment, anaerobic digestion is used to process solids. It stabilizes them using anaerobic microbes to degrade the organic waste to less objectionable, stable end products such as methane. Our guts do essentially an equivalent digestion process so that our bodies can dispose of waste. Describe how digestion in our bodies is similar or different to digestion in wastewater treatment. The speaker talked about the role of diverse microbial communities in anaerobic digestion, and how they can be influences by the introduction of chemicals such as triclosan and triclocarbon. How do you think our human guy microbial communities compare to wastewater digesters—e.g., more/less diverse, same/different microbes? Do you think our microbes are influences by ingestion of different foods and chemicals? Some researchers have looked at bioaugmentation with "good" microbes to help fix non-functioning digesters. Could we also do this if our gut biota isn't working efficiently? How?

Prompt 7: *Flint Water Crisis*

Over the past year, we have experienced one of the worst environmental engineering disasters in modern times in Flint, MI. In this case, the water source for the City of Flint was switched from Detroit water to the Flint River. The water quality of the river water differed from the previous source, and although the treatment plant had no problems treating the water to meet regulations, the chemistry of the new water (pH, corrosivity) caused lead to leach from the lead laterals in the drinking water distribution system. This caused enormous spikes in lead concentrations in some households (even to the point of being at toxic waste levels in some cases). The speaker referred to the water chemistry during his presentation, and also pointed out that lead is not a new problem. While we don't have thermodynamic data for low temperatures, we do have a lot of information about when lead will be soluble in solution. In the case of Flint, precautions such as adjusting the water pH or adding orthophosphate as a corrosion inhibitor were not employed. This leads to the other part of the story. The response by authorities in the case of flint was far from ideal, and some parties have been prosecuted as a result of negligence or even telling flat out lies or covering up evidence.

These events have led to many discussions of ethical conduct, both by engineers and others. What are you perspectives on Flint? Was this an engineering disaster? Who is responsible? Was it preventable? What can be done to prevent similar situations in the future? What knowledge/skills do engineers need to learn in school to address and prevent crises like this?

Example 5. Internet Searches and Discussion Prompts: Real-World Benefits of Implementing Technology into Manufacturing

Author: Dr. Lisa Bosman
Course: Computer-Integrated Manufacturing
Learning Activity Approach: Internet Searches and Discussion Prompts

Description

This course is an elective for Mechanical Engineering majors. It provides an overview of the philosophy behind computer-integrated manufacturing, the various components involved, and an understanding of how to upgrade conventional facilities to a technology-enhanced operation. In this course, online discussions are used to explore real-world applications of computer-integrated manufacturing with a focus on challenges and benefits.

Learning Objectives

1. Describe an understanding of topical areas associated with computer-integrated manufacturing.
2. Critique and explore the challenges and gains around implementing technology enhancements and innovations in the manufacturing sector.

Learning Activities

Online Discussion: Each week, students post an initial post and a response based on the discussion prompt. Example discussion prompts are provided.

Discussion Post 1: *Overview and Introduction to CIM*

Computer-integrated manufacturing (CIM) is an umbrella term for the total integration of product design and engineering, process planning, and manufacturing by means of complex computer systems. Less comprehensive computerized systems for production planning, inventory control, or scheduling are often Considered part of CIM. By using these powerful computer systems to integrate all phases of manufacturing, from initial customer order to final shipment, firms hope to increase productivity, improve quality, meet customer needs faster, and offer more flexibility. For example, McDonnell Douglas spent $10 million to introduce CIM in its Florida factory. The computer systems automatically schedule manufacturing tasks, keep track of labor, and send instructions to computer screens at workstations along the assembly line. Eliminating paperwork led to an increase of 30% in worker productivity. Less than 1% of U.S. manufacturing companies have approached full-scale use of CIM, but more than 40% are using one or more elements of CIM technology.

Part 1: Use the Internet to find an example, where a U.S. manufacturing company has incorporated one or more elements of CIM technology. Describe the company.

What product do they make? How is technology being incorporated? What is the cost of the technology? What are the benefits of incorporating the technology? Be sure to cite your source(s).
Part 2: What challenges currently face computer-integrated manufacturing? Be sure to cite your source(s).

Discussion Post 2: Automation

The popular press often writes about the factory of the future: a fully automated factory that manufactures a wide variety of products without human intervention. Although some "peopleless" factories do exist and others will be built, the major advances being made today occur in manufacturing operations where computers are being integrated into the process to help workers create high-quality products.
Part 1: Use the internet to find examples where automation is driving factor where (1) a fully automated factory exists and (2) computers are being integrated to help workers. For each example, describe the company. What product do they make? How is technology being incorporated? Be sure to cite your source(s).
Part 2: What are the keys to becoming the manufacturing facility of the future? Be sure to cite your source(s).

Discussion Post 3: Robotics

Through widespread use of robots, an automobile manufacturer improved its global competitiveness and economic success. Much of the savings resulted from reducing its workforce from 138,000 to 72,000. There was a human cost of displaced workers, however, and displaced employees had a difficult time finding new jobs. Was the automation decision defensible on ethical grounds? What steps can a firm take to be a responsible and ethical employer when cutbacks are necessary?

Part 1: Use the Internet to find an example, where a manufacturer was criticized for replacing workers with robots. What lessons can be learned from this example? Be sure to cite your source(s).
Part 2: What trends currently exist in computer-integrated manufacturing? Be sure to cite your source(s).

Discussion Post 4: RFID

Radio frequency identification (RFID) has been identified as one of the ten greatest contributory technologies of the twenty-first century. This technology has found a rapidly growing market, with global sales expected to top US $7 billion by 2008. An increasing variety of enterprises are employing RFID to improve their efficiency of operations and to gain a competitive advantage. RFID nodes can be mounted on various parts of machinery and plant to promote early fault detection and analysis. Their small size and autonomy enable their placement in locations that are usually difficult to access. In addition, it is also possible, with minimal changes to the machine configuration, to deploy sensors on the machinery after it has been installed. The sensor nodes not only monitor their own output but also collaborate with neighboring nodes to determine the health of the overall machines and provide early warnings of potential failure.

Part 1: Use the Internet to find an example where a manufacturer incorporated RFID as a means to monitor machinery. Describe the company and the benefits of deploying the technology. Be sure to cite your source(s).
Part 2: Identify a recent innovation in computer-integrated manufacturing. What is the value proposition? Be sure to cite your source(s).

Discussion Post 5: Lean Manufacturing

Waste elimination is one of the most effective ways to increase the profitability of any business. Processes either add value or waste to the production of a good or service. To eliminate waste, it is important to understand exactly what waste is and where it exists. While products significantly differ between factories, the typical wastes found in manufacturing environments are quite similar. The seven major wastes of a manufacturing environment include: (1) Production of defective parts, (2) Production of more parts than needed (overproduction), (3) Excessive inventories, (4) Unnecessary processing steps, (5) Unnecessary movement of people, (6) Unnecessary handling of materials, and (7) Workers waiting.
Part 1: Select one of the seven wastes identified above and use the Internet to find at least two commercially available technology-based solutions to measure the waste, decrease the waste, or eliminate the waste altogether.
Part 2: Respond to a peer's post. Comment on the potential pros and cons associated with each solution. Make a recommendation on which solution might be more feasible to implement first.

Discussion Post 6: Automated Assembly and Production Lines

The assembly line was one of the key components of the Industrial Revolution. Prior to assembly line production, items were often made one at a time by hand by a single crafter. This meant that there were often great variations between one crafter's work and the work of another crafter, and even among the products of a single crafter. If one part of a musket or tool were to break, it was no simple task to replace that part. Repairs and replacements had to be custom made to fit the specific item at hand. The principles of the assembly line allowed manufacturers to produce greatly increased amounts of products at lower cost and indirectly made for easier maintenance of products after their assembly. Many manufacturing operations have now upgraded to automated assembly and production lines. Control functions in automated assembly and production lines include (1) Sequence control—To coordinate the sequence of actions of the transfer system and workstations, (2) Safety monitoring—To avoid hazardous operation for workers and equipment, and (3) Quality control—To detect and possibly reject defective work units produced on the line.
Part 1: Use the Internet to find a manufacturer of automated assembly and production line applications (e.g., Rockwell Automation). Select a specific application and comment on how the manufacturer incorporates sequence control, safety monitoring, and quality control into the system design.
Part 2: Respond to a peer's post. Compare and contrast your own finds to your peer's findings. Are all manufacturing companies equal or do some offer a competitive advantage?

Discussion Post 7: Statistical Process Control

Statistical Process Control, commonly referred to as SPC, is a method for monitoring, controlling, and, ideally, improving a process through statistical analysis. The philosophy states that all processes exhibit intrinsic variation. However, sometimes processes exhibit excessive variation that produces undesirable or unpredictable results. SPC, in a manufacturing process optimization context, is used to reduce variation to achieve the best target value. The seven tools of SPC include Control charts, Histograms, Pareto charts, Check sheets, Defect concentration diagrams, Scatter diagrams, and Cause and effect diagrams.
Part 1: Use the Internet to find a commercially available technology application (e.g., software or computer application) incorporating one or more of the seven tools of SPC. What are the marketed benefits of the technology application? Provide an example of how this technology application could be used in the manufacturing environment.
Part 2: Respond to a peer's post. Assume you are the Director of Quality Control for a mid-size manufacturing company. What questions might you ask or research might you do prior to making a purchasing decision? Who are the stakeholders?

Discussion Post 8: Inspection

Properly designed and programmed automated inspection and gauging systems are very fast, accurate, and repeatable. They virtually eliminate the human error content of most gauge repeatability and reliability studies of manual inspection and gauging processes. Machine vision cameras can be programmed to automatically verify dozens of features on 100% of the products or assemblies produced in as little as one second. Precision laser sensors provide micron-level measurement accuracy without even contacting the part. Manual inspection, on the other hand, is highly labor intensive and error prone. Operators must rely on detailed work instructions, handheld measurement devices, and can be easily distracted while in the middle of checking the product.
Part 1: Use the Internet to find a manufacturer of automated inspection equipment. Identify its value proposition—In addition to the equipment, what other services/products/guarantees does it provide?
Part 2: Respond to a peer's post. Assume you are the VP of Research and Development for that particular Automated Inspection Manufacturing Company. Recommend how you might modify the value proposition to increase sales. What additional services/products/guarantees might you offer potential customers?

Example 6. Discussion Prompts to Respond, Compare and Contrast: The Customer Side of Environmental Chemistry

Author: Dr. Patrick McNamara
Course: Environmental Chemistry
Learning Activity Approach: Discussion Prompts to Respond, Compare and Contrast

Description

This course is a requirement for all Environmental Engineering majors. It includes many topics such as chemical stoichiometry, kinetics, and equilibrium, in addition to basic organic and inorganic chemistry concepts. Online discussions were used to help build interest and connection real-world applications to the area of environmental chemistry.

Learning Objectives

1. Develop students' ability to connect chemistry fundamentals to the world around them.
2. Critique and explore the real-world applications and challenges related to environmental chemistry.

Learning Activities

Prompt 1: *Introduction and General Chemistry*

Wisconsin's Department of Natural Resources is placing tighter restrictions on the effluent phosphorus levels in water to prevent algae blooms and the consequences associated with them.

Toledo has recently had problems with excessive water-borne phosphorus—http://www.toledoblade.com/local/2014/08/03/Water-crisis-grips-area.html—producing a significant algae bloom rendering the city's water supply undrinkable due to the microsystin microcystins and forcing hundreds of thousands of Toledo residents to use bottled water.

For this discussion, take a stand as to which approach you, as an environmental engineer, would recommend:

1. Increase the restrictions on effluent phosphorus levels to stave off potential algae blooms that may or may not happen, OR
2. Leave the phosphorus restrictions unchanged and have people use bottled water (remember, microsystins microcystins cannot be eliminated by boiling) should an algae bloom disrupt the water supply.

In your recommendation, take into consideration the short-term and long-term impacts, including costs, convenience, public sentiment, and the environmental/chemical considerations/consequences of each approach.

Finally, describe how you might deal with this situation, technologies you imagine developing, systems you might put in place if there was no budget limit.

In your responses to others, look for ways in which your position and perspectives are supported and refuted by others and for opportunities to compare and contrast your experience and position with others.

Prompt 2: Metal Complexation, Solubility of Salts, and Precipitation

Let's revisit the discussion about how to handle phosphorus. Now you know that one approach to managing phosphorus is adding iron to form solids that can be settled out during wastewater treatment.

Do you choose this treatment option or have everybody drink bottled water if we have an algae bloom? Why?

Are there other things we can add besides iron…what are they? (SOLUBILITY TABLE NEEDED FOR THIS) refer to the solubility table to answer this.

If you were hired as a consulting engineer would you recommend they switch away from iron? Why?

In your responses to others, look for ways in which your position and perspectives are supported and refuted by others and for opportunities to compare and contrast your experience and position with others.

Example 7. Ethics and Though Provoking Discussion Prompts: Software Medical Innovations and Issues

Author: Dr. Samuel Bechara
Course: Introduction to Computing for Biomedical Engineers
Learning Activity Approach: Ethics and Thought Provoking Discussion Prompts

Description

This course is a requirement for all Biomedical Engineering majors. It includes hands-on practical experience in programming, such as MATLAB, Solid Modeling, and CAD. Course topics include data types, syntax, control flow, and development of algorithms. Online discussions were used to promote creative thinking and understanding of Biocomputing challenges and issues in a real-world context.

Learning Objectives

1. Examine and discuss societal norms related to biomedical and modern technology.
2. Apply creative thinking to ethical dilemmas within the area of computer science.

Learning Activities

Discussion 1: Algorithms

As we learned in class from the phonebook example, algorithms are present in our everyday life and most of the time we don't even realize it. Your responsibility for this discussion post is to identify a unique algorithm that people use without even realizing it. The algorithm can be obscure (only a small subset of the population use it) or very widespread. It can be something that you use in your everyday life, something that people used to do in the past, or something they will do in the future.

I encourage you to think about several algorithms and choose the one that is the most unique and will be the most fun for your classmates to read. You can use multiple sources to search for ideas but do not just steal something that you find online.

After you have gone through the algorithm identification process outlined above, describe what your algorithm is accomplishing in one paragraph. You should also make a flowchart or write out the distinct steps using words. Be as descriptive as possible.

After you have submitted your algorithm to the appropriate discussion forum, read and critique at least one other students algorithm.

Discussion 2: Programming Logic

As we are learning in class, and as you already likely know, computers are just logic machines. We program them with whatever logic we decide and they will follow that logic to its conclusion regardless of the consequences.

The rigidity of computers following a predefined logic can clash especially with human ethics. One of my favorite thought experiments (and one of the most famous) is the "trolley problem".
The trolley problem has several variations but they all go something like this:

There is a runaway train barreling down the tracks. Ahead on the tracks, are 5 people that are tied up and unable to move out of the way. The train is barreling straight ahead and if it holds its course, all 5 of them are going to be killed. You are standing too far away to get to them in time to help, but are right next to a lever. If you pull the lever, the train will switch to another set of tracks. However, you notice that there is one person that is tied up on the side track.

Your options are:

(a) *Do nothing, the 5 people are killed*
(b) *Pull the lever, the one person is killed.*

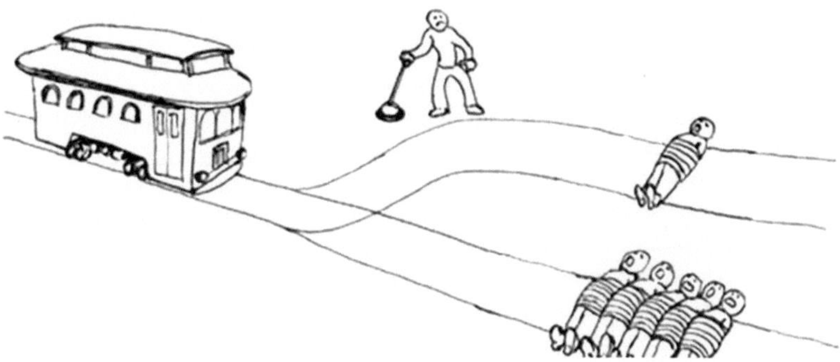

We are going to pretend that you are a distinguished programmer at Tesla and you are in charge of programming the autonomous driving function. You realize that you need to program what the car should do if it is in autonomous mode and a crash is inevitable in the scenario described above. You know that you can't write a program for EVERY possible combination of the trolley problem, so come up with an algorithm for the car to adhere to in the case of a trolley problem scenario. It is the future after all, so you can assume that you have access to whatever information about the 6 people involved as you like (age, weight, sex, religion, criminal history, job, etc.).

Describe what your algorithm is doing by writing out the distinct steps using words. Be as descriptive as possible. After you have submitted your algorithm to the appropriate discussion forum, read at least one other students algorithm and see if you can come up with a distasteful scenario for their algorithm.

Example: In the explanation of the trolley problem above. Let us say that Student A came up with an algorithm that simply values the number of humans involved. So my car would choose to swerve and kill the one person, instead of the five. The needs of the many outweigh the needs of the few and so forth and what not. So Student B's "distasteful" scenario would be one in which the one person was actually the man that tied the five up in the first place! So the car is going to spare his life and will kill the five innocent people! See how that sucks? Doesn't feel right does it?

Note: Your algorithms should be significantly more in depth, this just gives you an idea.

Discussion 3: *Correlation Versus Causation*

There is a very strong correlation between people who drowned after falling out of a fishing boat and the marriage rate in Kentucky. And if we are forced after the fact, we might even be able to come up with a reason that one causes the other. Maybe with less marriages, less women are pushing their husbands out of boats?

Ok, that is ridiculous, but what this is getting at is that correlation does not imply causation. This is a danger in data analytics and trend analysis. As we get better and better at analyzing data, we are going to find several correlations that are not this ridiculous but are equally as dubious.

Your job is search the Internet to find (1) an instance where correlation does not imply causation and (2) an instance where correlation does imply causation. In both cases, use logic to justify the causation in one paragraph.

Respond to at least one classmate, and either help build up their theory, offer an alternative to explain the causation, or offer an explanation as to how they cannot be related.

Discussion 4: *Software Innovation Ethics*

Radiation therapy is a lifesaving therapy that uses high energy radiation to kill tumors and cancer cells. That sounds all well and great but really radiation therapy sucks because the radiation is not specific to cancer. It damages EVERYTHING, but it kills cancer super hard (because it is rapidly undergoing mitosis). That being said it is much better than chemotherapy in terms of targeting disease sites.

The Therac-25 medical radiation device was involved in several cases where massive overdoses of radiation were administered to patients between 1985 and 1987. Be sure to read the article.

As a student, when you turn in software that doesn't work you get a bad grade. When the Therac-25 medical radiation device programmers turned over bad software, people died. For your discussion post this week you have a couple of prompts:

- Why do you think this happened?
- What do you think the punishment should have been for the company?
- Do you think that the individuals involved should have been punished?

Please respond to someone else's discussion post in a thoughtful and debatable manner.

Discussion 5: Technology Over Time

During the nineteenth century, a group of English textile workers began a movement to protest the advancements of technologies that made their skill (weaving) obsolete. They were called Luddites. (Note: Keep in mind we are referring to the nineteenth century movement, not the current term meaning someone opposed to technological change.)

For your post this week, put yourself in a nineteenth century textile worker's shoes. Your post should be a short paragraph as if you were writing a letter to your government representative arguing that the new technology is bad. Explain why. Respond to at least one other person's discussion post as the government official. Do you agree with their points? Do you have a vision for the future that they are missing? Both of your posts should be written in first person tone, so role-play a little bit.

Discussion 6: Autonomous Machines

We are in the midst of a rapidly changing world. Autonomous machines (and thinking machines) are capable of doing a lot of the work that was previously done by blue collar workers. Arguably, the political turmoil of 2016 could be seen as a product of disenfranchised middle class workers that are being squeezed out of the middle class by increasingly sophisticated technology.

There is a fantastic video about the current state of automatization. It is about 15 min and is very entertaining. Click here to check it out before you answer the discussion post (it may help to sketch out some notes): https://www.youtube.com/watch?v=7Pq-S557XQU.

Pick a side, in the video he argues that engineers and programmers make mechanical minds which in turn makes human brain labor less in demand. He convincingly argues that almost NO jobs are safe from computers.

The scope of this discussion post is *not* to agree or disagree with the video about whether automation is inevitable (it is). Your task is to come up with a solution to his conclusion. What do you think we should do when large sections of the population are not even employable (i.e., they are the horses in the video)?

Respond to at least one other person's post. Ask follow-up questions. Do you disagree or agree with them?

Discussion 7: Great Software Innovations

I hope that you all are with friends or family over this thanksgiving break and that your pie tin overfloweth. In the spirit of giving thanks, your task for the discussion post this week is to identify a piece of medical software or a specific algorithm that you are thankful for and explain why the innovative medical software is so great. Respond to at least one other person's discussion post.
Here is what I am thankful for:

I am thankful for computer-aided diagnosis algorithms for breast cancer. Although I have not personally known a woman that has been saved by these algorithms, I know that thousands of women have. Approximately 12.4% of all women in the US will eventually be diagnosed with breast cancer (source: https://seer.cancer.gov/statfacts/html/breast.html).

It is encouraging that with this technology, less diagnosis will be missed and more mothers, daughters, sisters, and friends will get to live longer and enjoy their lives. I am also impressed with the ingenuity of the programmers. Current algorithms have been shown to be successful in picking up breast abnormalities 87% of the time! (source: https://www.ncbi.nlm.nih.gov/pubmed/12091686).

I am glad that we live in a world where humans spend a significant amount of their brain capital to help other people.

Discussion 8: Predictive Software

Philip K Dick's short story Minority Report is one of my favorite short stories, go find a copy and read it for fun this winter.

The story is set in a future where three mutants can see all crime before it occurs. The "precogs" are plugged into a machine, and the Precrime Division arrests suspects before they inflict any harm. The ethical implications of this are obvious but this isn't something we have to worry about until the distant future, right?

Wrong! Humans proved we don't need mutants, we have computers and have developed "precog" algorithms already. The scary thing? They work!

Please read this article by the guardian before submitting your discussion post: https://www.theguardian.com/cities/2014/jun/25/predicting-crime-lapd-los-angeles-police-data-analysis-algorithm-minority-report

For your discussion post this week, respond to the following:

- What value does predictive policing offer?
- Who are the potential customer segments and stakeholders?
- Do you think it has the potential to be biased?
- Is it better than relying on human judgement or does it remove a human component that is important to policing?
- Are you concerned by any privacy concerns that arise from this technology (especially the data collection component)?
- What about the categorical suspicion that is mentioned?

Comment on at least one other person's discussion post. Remember to have a discussion. Be thoughtful and cordial in your responses.

Chapter 10
Example Engineering Curriculum: Value Creation Projects and Entrepreneurially Minded Tools

Author Reflection

No need to recreate the wheel…

At some point in our engineering career, most of us have received this great piece of wisdom: "Don't reinvent the wheel." This is commonly stated in reference to a proposed new design that incorporates a nonstandard size bolt, screw, or fastener. In designing for manufacturability, it is good to consider ANSI hardware design guide and charts.

The same wisdom can apply to developing learning activities. A learning activity incorporating the entrepreneurial mindset doesn't need to be made from scratch and require rocket science for it to be effective. Instead, there are many entrepreneurially minded tools, freely available online, which can be easily added to an EM learning activity.

This section provides an introduction and overview of curriculum ideas incorporating readily available entrepreneurially minded tools.

L. Bosman and S. Fernhaber, *Teaching the Entrepreneurial Mindset to Engineers*, DOI 10.1007/978-3-319-61412-0_10

Example 1. Four-Phase Semester Long Project: Biocomputing Real-World Design Application

Author: Dr. Taly Schmidt
Course: Biocomputing Design Lab 2
Learning Activity Approach: Four-Phase Semester Long Project

Description

This course is a requirement for all Biocomputing Biomedical Engineering majors and is a continuation of Biocomputing Design Lab 1. It focuses on medical devices utilizing embedded hardware and software. Since the course offers hands-on practice with software design and validation, the addition of a four-phase semester long Biocomputing investigation and design application offered a complimentary extension of the lab environment into current real-world medical technology applications.

Learning Objectives

1. Discuss trends and developments in the field of radiology software.
2. Question the current status, trends, and developments in the field of radiology software.
3. Critique and explore a contrarian view of currently available radiology solutions.
4. Create an innovation aimed at an opportunity which incorporates customer desirability, technology feasibility, and/or business viability.

Learning Activities

Radiologists provide approximately $1000 of value to the health care system for each hour that they work. Diagnostic radiologists spend the majority of their day interacting with software tools. Therefore, new software tools that increase a radiologist's efficiency, reduce errors, or improve comfort and job satisfaction will greatly reduce overall healthcare costs. This four-phase semester long scaffolded project offers a series of activities both in and outside of the classroom with a focus on the broad area of software tools for Radiology. Activities include identification and discussion of recent issues and products in the area of Radiology software, analysis and discussion of the unintended consequences and potential risks of radiology software, site visit, and interviews at a radiology department in a major hospital, development of a proposal for a new software product.

Phase 1 (6 Weeks): Online Discussion to Explore the Trends, Issues, and News Related to Radiology Software

This online discussion will take place each week over a period of 6 weeks. Each week, students post an article (news or scientific) related to trends, issues, and new developments in the field of radiology software. Each week students are expected to:

- Post an article (news or scientific) related to radiology software or the business of radiology that you find interesting.
- Read the articles posted by classmates and participate in online discussion.

Phase 2 (2 Weeks): Fieldtrip and Interview at Local Medical College

As a class, students visit the local Medical College to get further insight into radiology software trends and issues. Students come prepared with a list of questions to ask the engineers and radiologists.

Phase 3 (2 Weeks): Classroom Debate on the Computer-Aided Detection in Mammography

Students use a structured debate format, to answer the question of whether a radiology department should purchase and implement computer-aided detection (CAD) software tool for mammography. This CAD in mammography is a software tool that has been developed (and FDA approved) to potentially improve patient outcomes and radiology workflow. However, this software tool also has unintended negative consequences. The structured debate flows as follows:

- Position Presentation—Pro (5 min), Con (5 min)
- Work Period (5 min)
- Rebuttal—Pro (3 min), Con (3 min)
- Work Period (3 min)
- Response—Pro (2 min), Con (2 min)
- Work Period (1 min)
- Position Summary—Pro (2 min), Con (2 min).

Phase 4 (6 Weeks): Design Project to Identify an Unmet Need in Radiology Software and Propose a Solution

Students identify an existing problem in radiology software and to propose a solution to improve radiology outcomes or workflow. The proposed solution could be a new tool or a modification to an existing tool. Students present a 7 min pitch describing the problem, the proposed solution, potential risks, and the potential impact.

Example 2. Ethnography Driven Research: Innovation and Design of Cellular Telephone Systems

Author: Dr. Doug Melton
Course: Antenna Theory and Design
Learning Activity Approach: Ethnography Driven Research

Description

This course is a design elective for students obtaining a focus in Electromagnetic Fields and Communication. The course offers an introduction to diffraction theory, investigating the design and use of varying antennas including broad bands, horn, wire, and reflector antennas. Ethnography is introduced as a focus for innovation and design of a cellular telephone system.

Learning Objectives

1. Predict diffraction path-loss (in dB) for terrestrial wireless connections.
2. Demonstrate curiosity about the future of mobile wireless communications.
3. Express technical, societal, and economic insights regarding mobile communications.
4. Craft a value proposition for mobile wireless communications in a new context.

Learning Activities

These assignments allow the student to assume the role of an ethnographer and become more curious about wireless communications.

Assignment 1 (2 Weeks):

Create your personal set of questions and observations about wireless communication.

Assignment 2 (2 Weeks):

Create a set of questions for others about wireless usage habits, carriers, terrain, tower locations, and other factors that might influence propagation paths. Interview 10 people about their experiences with wireless communications.

Assignment 3 (2 Weeks):

Become an antenna aficionado. Take a photo of 20 mast-mounted antennas over the next week. In your antennae portfolio, accompany each photo with the GPS coordinates, estimated tower height, frequency, and modulation method. You may wish to verify your estimates with services such as www.antennasearch.com.

Assignment 4 (2 Weeks):

Predict trends in the following categories:

- Technical, e.g., "What will be the next major technical advance in mobile telecommunications?"
- Economic, e.g., "What seemingly unrelated business sector will be next affected by mobile communications?"
- Societal, e.g., "What are cultural changes (caused by wireless communications) that have just begun and are likely to continue?"

Assignment 5 (2 Weeks):

What trends, if any, are related to path-loss (distances, obstructions, multiband systems)? For example, consider T-Mobile's promotion of WiFi-connected telephone calls. These types of solutions reduce the reliance upon cellular towers, mitigating the large path-losses associated with spare tower placement. With the prevalence of WiFi, what alternative futures might exist for mobile communications?

Assignment 6 (2 Weeks):

Read the following articles and create a reflection on how the information in the articles is related to propagation path-loss. Record your three most valuable insights.

- Path-Loss Equations—Read the course text and provide comparisons of the author's treatment against two additional off-line resources.
- Newsletter for USAID's Nepal Economic Agriculture and Trade (NEAT) Project
- Deloitte report, "What is the impact of mobile telephony on economic growth?"
- WIRED Magazine article, "Where Cellular Networks Don't Exist, People Build Their Own"
- Mobile Technology Association of Michigan's eclectic list of "Unusual Uses of Mobile Technology"
- An article of your choice, unrelated to wireless communications.

Assignment 7 (2 Weeks):

Create an opportunity matrix for wireless communications. Form the rows of the matrix by making a list of twenty people groups (e.g., farmers, dentists, soccer fans, painters). Next, form the columns by identifying various wireless solutions, mobile platforms, technologies, features, (e.g., Snapchat, SMS, Peer-to-Peer, Satellite Phones). At each intersection in the matrix, identify how the specific group might benefit from the specific wireless solution.

Assignment 8 (2 Weeks):

Based on the opportunity matrix that was developed, select the most promising opportunity. Create a value proposition accompanied by a technical, societal, and economic defense.

Example 3. Five-Phase Semester Long Project: Free Open-Source Software (FOSS) Bugfix

Author: Dr. Dennis Brylow
Course: Topics in Computer Science—Software Engineering
Learning Activity Approach: Five-Phase Semester Long Project

Description

This course is an elective for Electrical and Computer Engineering majors. It focuses on software system design and implementation taking into consideration the potential challenges and technology. Since the course offers hands-on practice with design methodologies and patterns, configuration management, revision control, and validation, the FOSS Bugfix project is an ideal complement providing students the opportunity to explore real-world approaches to fixing software problems.

Learning Objectives

1. Evaluate opportunities and resources required to work on a FOSS bugfix.
2. Analyze and rank feature specifications for a proposed FOSS bugfix.
3. Develop testing criteria describing requirements and performance metrics associated with the FOSS bugfix.
4. Construct and implement a bugfix using agile development sprints.

Learning Activities

Phase 1—Initial Pitch:

Students select a FOSS (Free, Open-Source Software) project for class work, prepare an initial pitch (3 min/3 slides), and respond to key questions:

- What does it do?
- Who is its community of users?
- Who is its community of builders?

Class will debate and vote to select half the projects for the Detailed Pitch.

Phase 2—Detailed Pitch:

Students form teams, prepare a detailed persuasive pitch (6 min/6 slides) on the need to allocate resources to their selected project in comparison to alternatives, with necessary details. The student task is to persuade the class that allocating resources to the selected project will have more impact than the alternatives. Items appropriate for a Detailed Pitch:

- Big picture view of project
- Motivation for the project
- State of project
- Size of code base

- Size of community
- Release cycle
- Bug volume
- Tools.

Phase 3—Feature Ranking Specification:

Student teams must submit a report that helps the project manager (instructor) make practical decisions efficiently, builds the manager's confidence in the team's technical and communication abilities, and persuades the manager to accept the recommendation. The report should include the following:

(1) Criteria to facilitate and justify bug and enhancement ranking,
(2) Prioritized list of bugs and/or enhancements,
(3) Estimate of time/resources needed to fix each bug and implement each enhancement, and
(4) Recommended order in which the bugs and enhancements should be developed to meet customer needs.

Phase 4—Roadmap Specification:

Student teams must submit a report (3–5 pages) and presentation (10–15 min) that helps the project manager (instructor) understand that your team is capable of the following:

(a) Accurately describing the behavior of an implemented enhancement and a fixed bug,
(b) Accurately describing a test plan that verifies that enhancements have been correctly implemented and bug correctly fixed, and
(c) Effectively working together as a team.

The report and presentation should respond to the following questions:

(1) What types of tests will be used during the development process?
(2) What tests will be used to verify that each enhancement has been correctly implemented?
(3) What tests will be used to verify that each bug has been correctly fixed?

Phase 5—Implementation and Final Demonstration:

The Implementation phase consists of agile development sprints with regular class demonstrations, culminating in a Final Demonstration.

Example 4. Experiential Learning Design Challenge: Landfill Entrepreneurship for Sustainable Design

Authors: Dr. Elise Barrella and Dr. Bethany Brinkman
Course: Introduction to Engineering (and other first-year courses)
Learning Activity Approach: Experiential Learning Design Challenge

Description

Introduction to Engineering is a required first-year course intended to introduce students to overarching themes and culture in engineering and in the curriculum. Topics of coverage include professionalism, engineering and society, sustainable development, engineering fundamentals, systems approach in engineering problem solving, as well as creative problem solving practices and design. The module is intended to help students apply engineering design processes to a short, stakeholder-oriented project, and to analyze and evaluate products, processes, or systems based upon economic, environmental, societal, and technical characteristics.

Learning Objectives

1. Explain the principles of circular economy or cradle-to-cradle design.
2. Identify a local waste stream and describe its size and useful characteristics.
3. Develop a solution for local stakeholders/customers that uses the selected waste stream.
4. Present a value proposition for the proposed design (i.e., waste-to-wealth).

Learning Activities

The module is designed to take approximately two weeks, including both in-class and out-of-class work.

Background Reading for Discussion

1. McDonough, W., Braungart, M., Anastas, P.T., & Zimmerman, J.B. (2003). "Applying the Principles of Green Engineering to Cradle-to-Cradle Design." Environmental science & technology, pp. 434–441. http://www.c2cproducts.com/images/Enviro_Sci-Tech_article.pdf
2. Ellen MacArthur Foundation. (2017). Circular Economy Overview. https://www.ellenmacarthurfoundation.org/circular-economy/overview/concept

Design Challenge Prompt

- Assume you are an entrepreneur starting out with the goal of creating sustainable products, or those that are economically, socially, and environmentally responsible from cradle-to-grave. Describe how you might create or design a product that has a second (i.e., repurposed) life. To do this, you will need to consider current "streams" of trash, how those could be diverted, and how the new product could be created.

Design Challenge Steps

In small groups, complete the following steps. The Design Challenge is evaluated by a rubric using a variety of criteria including Communication, Teamwork, Feasibility, Justification, and Creativity.

- Step 1: The local landfill receives approximately 130,000 tons of waste (roughly 740,000 cyd) each year from people and businesses. Find a local waste stream. Conduct online searches, talk with local businesses/organizations, create a waste diary of everything you/your roommate throw away in a week, or call/visit the Landfill. "Deconstruct" or understand the waste stream—what are its characteristics? How much of it is available? What are limitations?, etc.
- Step 2: Brainstorm at least 10 ideas for how you could demake or remake that waste stream to meet a local need. Choose your best idea.
- Step 3: Develop a value proposition (using a canvas) to communicate how your idea meets a community need and creates value from waste. (Example Canvas: http://www.peterjthomson.com/2013/11/value-proposition-canvas/)
- Step 4: Present your solution (including drawings and/or proof-of-concept prototype) and value proposition to the class.
- Step 5: Reflect on your solution. How does/could your remake reflect principles of Circular Economy and Cradle-to-Cradle design? How might businesses benefit in a circular economy? The environment? The public? Connect back to the course readings.

Potential Variations

The Demake/Remake module is also appropriate for a first-year seminar or an interdisciplinary seminar focused on sustainability or design. It can be connected to a larger class project (i.e., demake/remake to address an assigned community need) or stand-alone.

Optional Follow-up Reading (for reflection and case study):

1. Mair, J., & Ganly, K. (2010). Social entrepreneurs: Innovating toward sustainability. In *State of the World 2010: Transforming Cultures: From Consumerism to Sustainability*. WW Norton & Co Inc.
2. The Natural Step. "Interface: The Journey of a Lifetime" (2013)
3. Rich, Nathaniel. "The Lawyer Who Became DuPont's Worst Nightmare" (2016) The New York Times Magazine. http://www.nytimes.com/2016/01/10/magazine/the-lawyer-who-became-duponts-worst-nightmare.html

Example 5. Industry Focused Design Project: Mastering the Product Development Process

Author: Dr. Ken Christensen
Course: Computer Engineering Senior Project
Learning Activity Approach: Industry Focused Design Project

Description

This course is a requirement for all Computer Engineering majors. The course covers a focused team-based design experience incorporating appropriate engineering standards and multiple realistic constraints. Projects are proposed by industry and/or other partners and are completed within a defined product development process. This semester long course project helps students to learn and master the product development process including requirements gathering, design including ideation, testing, and final delivery and deployment to a real customer.

Learning Objectives

1. Demonstrate collaboration, communication, and project management skills taking into consideration the perspectives and motivations of industry partners.
2. Apply a formal development process to complete a project
3. Synthesize information including requirements documentation, specification documentation, and test plan documentation.

Learning Activities

Having good support from industry is key to the success of this course. A good strategy is to approach companies who hire graduates from your program, with a primary focus on companies within driving distance of the campus so that the students can easily travel to the corporate site (and the industry sponsors can travel to campus). This course offers two main benefits to industry. First, industry partners are allotted a first look at the graduating class for recruiting. Second, industry obtains an opportunity to have a noncritical path "back burner" problem solved.

Due Week 3: Requirements Document

- Do you know what problem you are solving? Do you know what your customer wants? Can you describe to your customer, in your words, that you understand their need? You must clearly and crisply describe the project requirements. Requirements must be numbered so that they can be traced throughout the development project. Page limit is two pages.

Due Week 6: Specification/Design Document

- Do you know how to solve the problem and meet your requirements? This document should clearly specify the solution and include a design. A design could be a highlevel flowchart or a hardware block diagram. The document must include details on the methods what will be used to solve the problem. The reader must be convinced that you have a workable and feasible design. Page limit is two pages (but appendices are allowed).

Due Week 9: Prototype Demonstration

- This is a major checkpoint. You must have something up and running. A prototype is something you can show your customer and ask, "Is this what you wanted?" The customer must be able to see that all features are in place (but, perhaps not fully working). For software, a prototype would include all interfaces.

Due Week 10: Test Plan Document

- How do you know you met the requirements (validation)? How do you know that you implemented your specification and design correctly (verification)? Describe your test cases for validation and verification. A test case must include requirement or specification number that is covered, procedure to execute, and expected results. Good test cases should cover expected and unexpected inputs. No page limit, but should be short.

Due Week 13: Practice Presentation

- This is a dry run of your presentation and demonstration. Your presentation should not, in any case, exceed 30 min. Expect lots of feedback and budget time to rework your presentation overheads for the final presentation.

Due Week 15: Final Submission

- Your final presentation and demonstration is given to your company supporter at the company site. You will deliver your final submission at this time. Your final submission should contain revised versions of all of the above documents, your presentation overheads, your artifact, and any final documentation that is required (e.g., user documentation if a software product, complete specifications if a hardware product). Your final submission also includes a poster and a press release.
- Poster—The poster is a self-explaining presentation of your project. The poster will be hung in an engineering design classroom for at least one semester. The organization of a poster will be discussed in lecture.
- Press Release—The press release is a one-page project description suitable for the general public. Unlike the other deliverables, the press release may be somewhat self-promoting.

Project Examples

Examples of four projects that show the diversity of the companies, types of projects, and student project results are shown here:

- Breed Technologies: Breed develops and manufactures products for improving vehicle safety; in particular, air bag related technologies. Breed had a group of students set up an Intranet web site for the Breed R&D department. The contents of the web site were the general information of the R&D department, a listing of the core team members, description of the core projects, and the department's Labor Tracking & Report System (LTRS). The LTRS system tracked the labor rates, materials procurement, outside contracts, and fiscal budgets for Breed internal R&D projects. Students designed the LTRS and wrote the software (using Microsoft ASP) for the website. The students also organized the information on the website to be useful to Breed internal needs. The students' work is in daily use by Breed.

- Catalina Marketing: Catalina develops targeted marketing services that use technology and information to reach consumers through many different media. One service that Catalina produces is the ability for cash registers to track consumer purchases and print appropriate coupons on the back of sales receipts. Catalina had a group of students explore solutions for compression of error logs to reduce dial-up line costs (e.g., from a failing cash register to the central Catalina headquarters). The desired solution was a C++ class that could easily be incorporated into existing Catalina software. Since the content of the error logs could be partially predicted, an innovative dictionary-based compression scheme could be implemented where the dictionary existed at both ends (and need not be transmitted as part of the compressed file). The status of the students' work is not known.
- Sonny's Bar-B-Q: Sonny's is the largest Bar-B-Q franchise in the country with over 130 restaurants. Sonny's is always exploring new applications of computing technology to the restaurant industry. Sonny's has contributed projects to (1) explore the use of handheld devices (e.g., Visor and Palm PDA's) for taking orders at the table and with a wireless network send the orders to the kitchen and (2) explore the use of handheld devices for barcode inventorying of food products to reduce loss. Students designed and wrote software for the handheld PDAs to achieve the goals set by Sonny's. This work was intended as proof-of-concept and thus was not directly implemented by Sonny's.
- Sypris Electronics LLC: Sypris is an electronics manufacturing and engineering services company serving the Department of Defense and the avionics and aerospace communities. Sypris asked a group of students to explore new methods of generating user passwords (for access to desktop computers and file servers). The generated passwords needed to be resistant to cracking by programs such as L0phtCrack, but also be memorable by a human. A user-friendly front-end for generating passwords was requested by Sypris. This project did not achieve very good results. The students produced a method that was not evaluated (and had potential holes for cracking of passwords) and hence was really not usable by Sypris.

Example 6. Gaining Empathy Through Avatars: A Thermodynamics Project

Author: Dr. Casey Allen
Course: Thermodynamics II
Learning Activity Approach: Gaining Empathy Through Avatars

Description

This course applies thermodynamics, heat transfer, and fluids concepts to combustion processes, power and refrigeration cycles, and psychrometrics systems. The lab session offers an ideal opportunity to incorporate the entrepreneurial mindset using avatars to simulate real-life interactions to gain empathy about design needs. The avatars were developed using Voki (www.voki.com).

Learning Objectives

1. Connect theoretical topics to real-world or technology-based applications.
2. Apply thermodynamic analysis to evaluate the technical and financial viability of engineering systems.

Learning Activities

Phase 1: Design Thinking Lab

You have been hired to design efficient processes that utilize waste streams from within and across five facilities for enhanced efficiencies. Identify opportunities to reuse waste energy/mass streams across/within these facilities (natural gas power plant, meat packing facility, steel mill, wastewater treatment facility, and trash incinerator). Complete the following design thinking tasks:

A. Empathize—Watch the five Avatar video interviews, formulate questions for the avatars, and document assumed answers with justification.
B. Define—State a specific challenge or problem faced by one or more of the avatars.
C. Ideate—Brainstorm a minimum of three ideas for addressing the challenge or problem. Present the ideas to another group to receive feedback (simulated customer feedback).
D. Prototype and Test—Conduct analysis on the select solution both in the lab and after to test the feasibility of the solution.

About the Avatar Videos

<table>
<tr>
<td>
</td>
<td>
</td>
<td>
</td>
</tr>
<tr>
<td>a. Energy/Mass Rejected: hot combustion exhaust from gas turbine
b. Energy/Mass Inputs: gaseous fuel (natural gas currently used)</td>
<td>a. Energy/Mass Rejected: heat from large -scale refrigeration system for meat coolers
b. Energy/Mass Inputs: electrical energy used to drive compressors in refrigeration cycle</td>
<td>a. Energy/Mass Rejected: blast furnace heat loss, exhaust gases (~20% CO)
b. Energy/ Mass Inputs: coke (fuel) for blast furnace, blast furnace cooling water</td>
</tr>
</table>

<table>
<tr>
<td>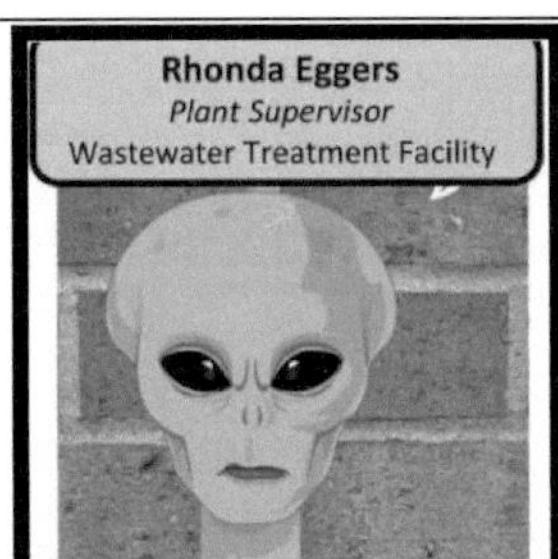
</td>
<td>
</td>
</tr>
<tr>
<td>a. Energy/Mass Rejected: biosolids (solid sediment), processed water (released to environment)
b. Energy/Mass Inputs: waste water, electrical power</td>
<td>a. Energy/Mass Rejected: hot combustion exhaust from industrial furnace, ash residue
b. Energy/Mass Inputs: gaseous fuel for co-firing burner (intermittent)</td>
</tr>
</table>

Phase 2: Report and Presentation Requirements

1. Introduction: (a) background information and motivation, (b) summary information gathered from interviews, including additional questions and assumed answers
2. Problem Definition: (a) statement of the problem leveraging insight from interviews, (b) evidence that supports the problem definition, (c) description of assumptions and interpretations

3. Selected Design: (a) description of the design process and why the final design was chosen, (b) technical description of the solution and rationale for how it solves the interviewee's problem, (c) hand sketches illustrating the selected design
4. Evaluation of Solution: (a) energy, mass balance, and efficiency calculations, (b) discussion of calculations and results for the selected design, (c) high level cost benefit analysis based on internet research, (d) discussion of issues that would still need to be addressed or questions answered to determine if the solution is viable, (e) comparison of proposed solution to technologies you find online.

Example 7. Design Thinking and the Value Proposition Canvas: Hypothesis Testing Through Statistical Analysis

Author: Dr. Lisa Bosman
Course: Probability and Statistics for Engineers
Learning Activity Approach: Design Thinking and the Value Proposition Canvas

Description

`This course is required for all Engineering majors. Statistics is the field of study focused on data collection, organization, analysis, and interpretation. This course covers topics such as descriptive statistics, data collection techniques, probability, graphs, and hypothesis testing. Since statistics can be used to validate assumptions and test criteria, it offers a great opportunity to explore and validate potential innovation ideas through data collection and analysis.

Learning Objectives

1. Describe basic statistical analysis calculations and interpretations of hypothesis testing related to observational study, surveys, and experimental design.
2. Understand the role of statistical analysis during the design prototype, testing, and validation process.
3. Create an innovation aimed at an opportunity which incorporates customer desirability, technology feasibility, and/or business viability.

Learning Activities

Phase 1 (Wks 1–3): Customer Discovery, Design Thinking, and the Value Proposition

Students are required to complete a 90 min (plus debrief) in-class design thinking project using example challenges as follows:

(1) Improve the Gift Giving Process
(2) Design New Wearable Fitness Tech
(3) Create the Perfect Carrying Bag, Purse, or Backpack
(4) Develop a Better Classroom Learning Experience
(5) Other.

The 90 min project allows students to go through all components of the design thinking process including Empathize, Define, Ideate, Prototype, and Test, resulting in an innovation idea and potential customer segment. Stanford University's d. School (http://dschool.stanford.edu/dgift/) has plenty of resources (e.g., lesson plans, worksheets, tutorials, etc.…) to get faculty started on the design thinking process.

Next, students will dig deeper into the idea by considering the value proposition. The Value Proposition describes how value is created for the customer. The Value Proposition Canvas visually shows how value is created for the customer. It ensures the gains and pains of the customer jobs (customer needs) are identified, and solved by a specific product or service that relieves pain or creates gain, while taking into consideration current solutions, and clearly explaining the value proposition summary (main product or service differentiators).

A tool for developing the Value Proposition Canvas is provided below. This was modified based on the Value Proposition Canvas available at https://strategyzer.com/canvas/value-proposition-canvas.

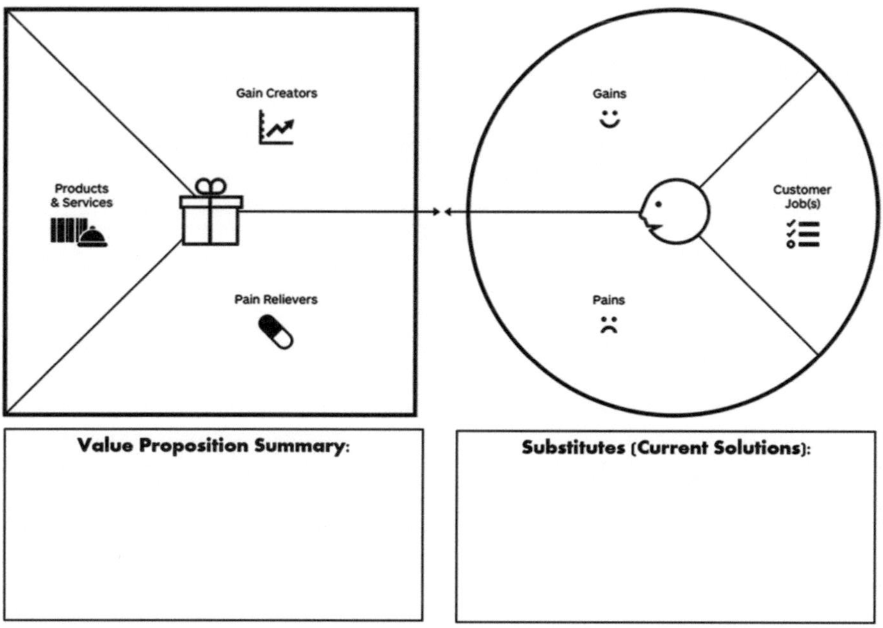

Phase 2 (Wks 4–6): Observational Study to Test Hypotheses

Students need to develop and deploy an observational study to test a hypothesis based on the Value Proposition Canvas (see Strategyzer Test Cards for more information).

For example, using the challenge "Design New Wearable Fitness Tech" a student came up with an innovation idea that uses a Smartphone and heart rate monitory to automatically keep track of exercising at the gym. The student assumed that the majority of people (51%) keep a Smartphone with them at all times, even at the gym.

Using this example, one possible observational study would be to observe and evaluate the proportion of people carrying Smartphones while exercising at the local gym. Students would collect data and the hypothesis test would compare the null hypothesis (H0 = 51%) to the alternative hypothesis (H1 < 51%).

Phase 3 (Wks 7–9): Survey to Test Hypotheses

Students need to develop and deploy a survey to test a hypothesis based on the Value Proposition Canvas (see Strategyzer Test Cards for more information). For example, students might send out a survey to friends and family using social media, or students might stand outside of the local gym and ask questions face-to-face.

Phase 4 (Wks 10–12): Experiment to Test Hypotheses

Students need to develop and deploy an experiment to test a hypothesis based on the Value Proposition Canvas (see Strategyzer Test Cards for more information). For example, students might randomly split a group of friends and family into a Control Group and a Treatment Group. The Treatment Group will receive feedback and data analytics while exercising and the Control Group will receive no information. A comparison could be done on the quantity of minutes spent exercise, rate of exertion, or many other variables associated with exercise and fitness.

Phase 5 (Wks 13–15): Formal Business Pitch

Students need to summarize findings, collect additional data related to the Business Model Canvas (based on the 2010 book developed by Alexander Osterwalder and Yves Pigneur), and formulate a business pitch considering the following elements:

- Value Proposition—What is the bundle of products/services being delivered? What customer problem is being solved? What customer need is being satisfied? What is the minimally viable product/service?
- Customer Segments—Who are the customers? Do multiple customer segments exist? If so, which is the most important? What is the customer archetype?
- Revenue Streams—What is the revenue model? How much are customers willing to pay? What do customers currently pay for substitutes?
- Customer Relationships—How is customer feedback integrated into the business model? How much does each customer spend? How much does each customer cost to market? How will you communicate with customers?
- Channels—How will you reach/interface with customers for marketing, purchasing, and delivery? How do substitutes interface with customers? Would customers prefer to be reached a different way?
- Key Partners—What key partners are required to produce and deliver the product/service? Who are the key suppliers? What are the major resources that need to be acquired from partners?
- Key Resources—What key resources are required to produce and deliver the product/service? What distribution channel will need to be developed? What technology will need to be developed to build customer relationships and obtain feedback?
- Key Activities—What key activities are required to produce and deliver the product/service? What distribution channel will need to be developed? What technology will need to be developed to build customer relationships and obtain feedback?
- Cost Structure—How much do key activities cost? How much do key resources cost? What are the major costs required? How much money is needed to get started?

Example 8. Student Select Design Projects: A Heat Transfer Challenge

Author: Dr. Anthony Bowman
Course: Fundamentals of Heat Transfer
Learning Activity Approach: Student Select Design Projects

Description

This course is a requirement for all Mechanical Engineering majors and offers an overview of theoretical and applied heat transfer including conduction, convection, and thermal radiation. The addition of a design component provides students the opportunity to connect academic theory to real-life applications.

Learning Objectives

1. Develop student ability to work effectively in teams.
2. Develop student written technical communication skills.
3. Design a system, component, or process integrating the three areas of thermal sciences—thermodynamics, fluid mechanics, and heat transfer.

Learning Activities

Phase 1: Team and Project Definition

Students are required to form teams, select a project, and submit a scope of work including parameters to investigate. See the list at the end for potential design projects.

Phase 2: Interim Progress Report #1

Students are required to submit a progress report including background research, literature review, design approach, and preliminary design.

Phase 3: Interim Progress Report #2

Students are required to submit a progress report showcasing a refined design and engineering calculations.

Phase 4: Final Report and Presentation

The Final Presentation will summarize the report and take into consideration organization, efficiency, style, and mechanics. The Final Report will focus on the following areas:

1. Technical Feasibility
   - Project Statement
   - Metrics
   - Objective Tree
   - Function Tree
   - Selection Matrix
   - Design Configuration
   - Engineering Analysis
2. Business Viability
   - Bill of Materials
   - Economic Analysis
3. Customer Desirability
   - Customer Segment(s) and Market Size
   - Investigation of Socially/Culturally Appropriateness.

Potential Design Projects

1. Trombe Wall: Design a wall system capable of supplying 75% of the heating load for a residential home (1300 ft^2) in Logan, UT. Use January conditions as the baseline.
2. Hydrogen Storage Tank: Design a lightweight onboard storage tank (60L) for cryogenic hydrogen (25 MPa, 80 K) that can withstand exposure to a diesel-fueled fire for 30 min without rupturing.
3. Exhaust System: Design a low-cost exhaust system to remove wood smoke from a small residential home in a developing country. The system must operate off-the-grid and use only components and parts locally available.
4. Compost Bin: Design a compost bin capable of operating year-round outside in the Milwaukee area.
5. Greenhouse: Design a self-sustaining greenhouse that can effectively operate year-round in the Milwaukee area. Determine cost, efficiency, and capacity of the design.
6. Evaporative Refrigerator: Design a residential refrigerator that works on the concept of evaporative cooling. The device should not consume any electricity. Consider construction and water consumption costs as well as site selection. Compute energy/cost savings, carbon credits, and cooling capacity.
7. Off-the-grid Ice Maker: Design an ice maker capable of freezing water without electricity or other infrastructure inputs, i.e., it should be operable only with human, animal, or other renewable inputs and cost effective for developing country applications.

8. Off-the-grid Incubator: Design an alternative powered incubator capable of utilization in remote off-the-grid locations. Your design must maintain a temperature of 40 °C for 24 h and be cheap for application to poor communities.
9. Modeling and Redesign of Cooling for an Eddy-Current Dynamometer: Create a thermal model for the current cooling system and create design changes to better cool the dynamometer at the present power rating and/or suggestions to extend the power rating within the current frame.
10. Design the Packed Bed for a Pebble Bed Nuclear Reactor (PBNR): Create a design for the size of the packed bed and size of pebble taking into account the heat transfer and pressure drop through the packed bed and its overall effect on the closed loop Brayton Cycle that the PBNR is based upon. The project will require a parametric analysis of the two parameters on overall net cycle efficiency and net power output of a 100 MW-based design.
11. Air-to-Air Heat Exchanger: Design an air-to-air heat exchanger (i.e., charge air cooler) for an over-the-road diesel engine application that will meet current Tier 3 emissions and future Tier 4 emissions requirements. The designs must fit within specified maximum physical dimensions and cost expectations.
12. Design of a Solar Pool Heater: Project includes engineering design and economic analysis (cost of materials, operation, and return on investment) of conventional systems to be compared with solar pool heater design. Heat loss/gain calculations to include solar gain during the day and radiation losses overnight. Solar pool heating system to be designed for two locations (Milwaukee area in late May and Dallas area in late February).
13. Modeling of Closed Loop Liquid CPU Cooler: Design a closed loop CPU cooling system maximizing heat transfer to maintain CPU at reasonable operating temperatures.
14. Waste Heat Recovery System: Design a waste heat recovery system and application for a Caterpillar hydraulic pump testing facility located in South Milwaukee. The test bed has the capacity to test up to a 1000 hp hydraulic pump. The pump is currently loaded by imposing a large pressure drop through a valve that is also used to vary the load per test specifications. The hydraulic oil absorbs the 1000 hp and as such must be cooled. The cooling is currently accomplished using city water and a shell-in-tube heat exchanger. The water bill for a month long test can exceed $15,000.
15. Blast Freezer System: Design a blast freezer system for use by UPS/FedEx to freeze PCM (Phase Change Material) packets used to keep pharmaceutical packages at −50 °C during shipment. The project requires an analysis of the blast freezer, PCM material, packaging/disposal, and freezing process. Additionally, the project team (or part of) may need to travel to the sponsoring company in southeastern Ohio during the Spring Break.

16. Solar Powered Cogeneration System: Design a solar powered cogeneration system to supply 3 kW of cooling for refrigeration or air-conditioning and store excess electricity as part of a larger green power distribution system that may include a wind turbine. The solar collector will include a concentrator and simple tracking mechanism. This project may be divided into sub-projects to be worked on by multiple groups.
17. Optimization of High Power Density Electronics Cooling (Project I): Model and optimize the maximum rate of heat transfer via convection for power electronics, subject to reasonable operating conditions and power duty cycle.
18. Optimization of High Power Density Electronics Cooling (Project II): Model and optimize the maximum rate of heat transfer using a chilled water loop for power electronics (collaboration with a local induction heat treating equipment company; NDA required), subject to reasonable operating conditions and power duty cycle.
19. Design of an R600- or R290-based (Butane or Propane) Refrigeration System. Design a modern refrigeration system to maximize performance (i.e., cooling capacity and power consumption) using currently available components, subject to target cold storage temperature and reasonable environmental operating conditions. The design challenge is in maximizing the cooling capacity given the restrictions on the maximum amount of refrigerant in the system, 150 grams (about 1/5th of the typical system).

Example 9. Constraint Focused Design Challenge: Fluid Mechanics and Redesign of a Resort Lodge Water Supply

Author: Dr. Andrew Gerhart
Course: Fluid Mechanics
Learning Activity Approach: Constraint Focused Design Challenge

Description

This course covers the basic principles and applications of fluid statics and fluid dynamics. It is a required course for Mechanical, Civil, Architectural, and Chemical engineers since it studies commonly encountered concepts of fluid continuity, flow work, flow momentum, and energy interactions of fluid systems. The learning activity helps students design a fluid delivery system including pipes, pumps, and the various components typical of such systems. The students will also need to engage with a customer, be cost conscience, and identify opportunities to create value.

Learning Objectives

1. Design a pipe system and choose an appropriately sized pump for the given fluid needs.
2. Distinguish the difference between pipe component options and pump options to minimize fabrication and operational cost.
3. Identify an alternate design (i.e., an opportunity) which will create higher value over traditional design.

Learning Activities

This is a four-week-long design project wherein the students work in teams. The opportunity is "hidden" with the considerations given as bullet points. Because the hillside continues steeply above the lodge/cottage, a tank can be placed at the top. A small pump at the lake can refill the tank at night when electricity rates are lower. During daytime water use, the entire system is gravity fed. Using this opportunity, the system does not need upgraded later. A sufficiently sized tank can supply the entire hotel needs.

Design Project Requirements: Wilderness Water System

Your rich uncle, Mortimer, has recently purchased a large tract of land in the Upper Peninsula of Michigan. He did not become wealthy by purchasing worthless things, yet the land he bought has no valuable minerals, nor any profit from lumber. It does have a magnificent wilderness resort lodge, but it was abandoned years ago and fell into a dilapidated state. The lodge is known as the Overlook Hotel. (No, not *that* Overlook Hotel from *The Shining*; that place makes people go crazy and is located in the mountains of Colorado.)

Before Uncle Mortimer can begin restoration of the Hotel, he needs a modest cottage for multiday stays while he begins planning. Besides a living room and bedroom, the cottage will have a kitchen with a sink and a bathroom with a shower, sink, and toilet. He also wants a spigot on the outside of the cottage for a hose to rinse the dust from his 2014 Ford F-150 SVT Raptor (yes, the plush one with a 6.2L V8 engine). The land has access to electricity. His cottage (and Hotel) will be on a rocky hillside 300 vertical feet above the lake (which is what the hotel will "overlook") and 2200 feet from the lake's edge. A water well cannot be drilled through the rocky hill.

After learning of your vast new knowledge of fluid mechanics, he has asked you to design a water system for his cottage. Eventually that same water system will be upgraded to supply the hotel. Your focus should be on the cottage's water system, while keeping in mind that the system will be enlarged in the future. You will need to consider a water delivery system, filter(s), heater(s), a piping system, and other components for this cottage. You must keep in mind that Uncle Mortimer is miserly with his expenses; he did not get rich by wasting money. But Uncle Mortimer is very generous with his family. Therefore, if you can design an efficient and cost effective system, you will not only be paid well, you will likely inherit the land and hotel in Uncle Mortimer's will!

Preliminary Reply Investigation: some (not all) considerations during the first week:

- How should water be removed from the lake?
- How much water is needed?
- What are some needs and consequences for Uncle Mort and his customers?
- What major components are needed for approximated water capacity?

Some considerations:

- Ensure that the cottage has typical/sufficient water flow and pressure.
- Be careful with pipe selection (sizing) and material, ensuring that the water is fairly equally distributed throughout the cottage.
- The layout of the water system will determine the layout of the cottage.
- Be cautious that the components and design are not too costly. You should keep track of approximate expenses for components. You do not need to consider installation costs.
- Consider operational expenses for Uncle Mortimer. In other words, choose your water delivery system and pipe layout wisely.
- The hillside continues above the cottage/hotel another 400 vertical feet to the summit in 600 ground feet.
- Consider Upper Peninsula weather conditions.
- You do not need to be concerned with sewage (i.e., sinks, shower, and toilet drains).
- Remember that this system will be upgraded for the entire resort lodge.

Potential Variations

Many variations can be made from this theme specifically or more generally. Specifically, a snow making system can be designed for the hillside, a sprinkler system, or a fountain system that does not need potable water.

Generally, pipe/pump systems can be the theme for opportunity recognition in other ways. One scenario has Uncle Mortimer supplying water on his manmade island off the coast of Dubai. A pipe/pump system with a desalination unit is the standard design option; the opportunity lies in the fact that potable water can be pumped directly from the Dubai city water supply 2 miles away. In another scenario, Uncle Mortimer is having a 20' by 12' salt–water aquarium installed in the main floor of his mansion with a small aquarium installed directly above the large tank in his office. Many students try to design a single pipe/pump system for both aquariums. This is costly; two pumps appropriately sized reduce the cost and complexity drastically.

Finally, the rich uncle theme can be replaced with water delivery systems for remote communities or for a large industry. The key is to have a customer/stakeholder.

Example 10. Service Learning and Design: A Multidisciplinary Approach to Assistive Device Design

Author: Dr. Cristi Bell-Huff and Heidi Morano
Course: Entrepreneurial Engineering Design Studio
Learning Activity Approach: Service Learning and Design

Description

This course emphasizes creating solutions through team-based projects utilizing engineering tools and skills, along with opportunity identification, ideation, value analysis, and customer engagement. In this sophomore level course, students build upon the foundation of entrepreneurial mindset development laid in the first-year introductory design course. At the sophomore level, students spend a semester on a multidisciplinary team that identifies opportunities for design within a theme, engages real customers beyond the walls of the classroom, and develop a working prototype that creates value for these customers.

The design theme for the studio, "Accessibility in the Workplace," is based on a partnership with a nonprofit organization that helps people with a wide range of disabilities develop, train, and place into meaningful employment. The students are tasked with designing an assistive device to help their customer perform their job. This real customer interaction fosters empathetic design while providing a more meaningful classroom design experience as students are able to see the direct impact their designs have in creating real value—value as it is defined, not by the student or faculty member, but by their customer.

Learning Objectives

1. Explain empathy for real customers through an Accessibility Simulation exercise.
2. Create value for customers through outcome-driven innovation.

Learning Activities

1. *Accessibility Simulation Exercise*

Before we introduce the "Accessibility in the Workplace" design theme, to set the stage and cultivate a user-centered, empathy-in-design approach to recognizing and addressing opportunities, we designed this simulation exercise to help students understand the everyday experiences of people with disabilities.

In this exercise, students will:

1. Select from five different simulation activities
    - Mobilitiy impairment (wheelchair or walker)
    - Dexterity impairment (hand braces that restrict range of motion of fingers)
    - Vision impairment (Low vision simulation kit)

- Hearing impairment (Lip reading activities)
- Autism simulation activity

2. Predict the most difficult aspects of dealing with the disability.
3. Perform a given set of common, everyday tasks.
4. Reflect on their experiences.

   - Did your predictions match your experiences?
   - How did it feel dealing with the disability?
   - What kinds of things did you have to go through to "accommodate" this disability?
   - Can you predict what specific issues someone with this disability might encounter in the workplace?

2. *Outcome-Driven Innovation*

The student teams interact with real customers at their workplace at *five* scheduled visits over the course of the semester:

1. Opportunity Identification
2. Customer Needs/Design Requirements
3. Customer Feedback
4. Prototype Testing
5. Delivery of Working Prototype

Getting the right customer input at these site visits are essential to the creation of value and to the overall success of their projects.

3. *Site Visit #2*

At Site Visit #2, the teams are tasked with developing a set of design requirements based on their customers' needs. To help focus their interactions we use an outcome-driven innovation approach:

- Ulwick, Anthony W., *What Customers Want, Using Outcome-Driven Innovation to Create Breakthrough Products and Services.* New York: *M*cGraw-Hill, 2005.

This approach has the students *focus on the job* as the unit of analysis. Based on their customer input, the student teams perform each of the following steps:

1. Deconstruct the job into steps, from start to finish.
2. Determine how success is measured at each step.

   For each step in the process consider:

   (a) What makes this step slow?
   (b) What makes this step inconsistent or unpredictable?
   (c) What makes this step costly or wasteful?

3. Translate each of these metrics into Customer Needs Statements that take the form:

   DIRECTION ... Unit of Measure ... Outcome Desired*

   - Note: *DIRECTION* must be expressed as either *"INCREASE"* or *"MINIMIZE"* (e.g., INCREASE the number of parts that can be concurrently assembled.)

4. Translate these Customer Needs Statements into Design Requirements that fall into one of three categories: (a) MUST DO, (b) MIGHT DO, (c) MUST *NOT* DO

Innovation must better satisfy the needs related to the customer getting the job done. We emphasize that its the job's perfect execution that reflects the customers' true definition of value.

Example 11. Applying the Business Model Canvas: Ideation and Biomechanics

Author: Dr. David Jamison
Course: Introduction to Biomechanics
Learning Activity Approach: Applying the Business Model Canvas

Description

The purpose of this course is to provide an introductory survey of topics from the field of biomechanics, including orthopedics (bone, soft tissue, joints), cellular mechanics, biomaterials, biofluids, and medical devices. The purpose of the project is to develop and present a profitable business idea that addresses an unmet need within the biomechanics field.

Learning Objectives

1. Identify an opportunity to investigate the market and test concepts quickly via customer engagement.
2. Create a preliminary business plan.
3. Communicate engineering solution in terms of economics and societal benefits.

Learning Activities

We have discussed various challenges within the biomechanics field—areas that can greatly benefit from innovative advancements. As a future biomechanists, you will have the necessary capabilities to use your technical mastery of biomechanics to develop innovative, impactful solutions to our current problems. Your objective for this term project is to work in teams to develop and present a profitable business idea that addresses an unmet need within the biomechanics field. Your business idea may provide a product or service.

Students will work in groups (2–3 students) to present a business idea relevant to the biomechanics field. Each group will deliver a business model canvas along with a pitch presentation. The presentation time for each group will be 12 min followed by 3 min for questions. Deliverables will be evaluated on the group's ability to explain the unmet need; articulate the relevant solution; and identify the business opportunities, costs, revenue streams, customers, added value, and societal impact.

For both deliverables (the business model canvas and pitch presentation), students will be evaluated on their ability to convey the following:

- The Business Idea (Product/Service): Students must clearly identify the customer, including who they are, what type of relationships will be establish with them, and how you will reach them (channels). In addition, students must communicate the value proposition in terms of economic, customer, and societal value.
- Market Research: Students must identify the unmet need that the business idea/solution satisfies. In addition, students must be able to test concepts via customer engagement.
- Costs and Revenue: Students must clearly identify cost structure, revenue streams, key resources (e.g., infrastructure), partnerships, and activities necessary.

Example 12. Weekly Deliverables for Semester Long Project: System Engineering Approach to a Capstone Course

Author: Dr. John Santiago and Dr. Jing Guo
Course: Capstone Design
Learning Activity Approach: Weekly Deliverables for Semester Long Project

Description

This course is a requirement for all masters' degree seeking Electrical and Computer Engineering majors. The capstone course provides the student an opportunity to integrate skills developed throughout the Master's program by completing a project or study that focuses on a technical problem or current issue in engineering. The students must: define the problem or opportunity, identify constraints, complete an analysis, prepare and deliver a professional report, and prepare and deliver a professional presentation.

Learning Objectives

1. Frame an ill-defined engineering problem and define appropriate requirements that the solution must satisfy.
2. Design and implement a test protocol to verify that the designed system meets the defined requirements.
3. Design and/or model a system to meet the defined requirements.
4. Integrate skills, concepts, and/or tools acquired in prerequisite coursework and apply the same to successfully complete the project.
5. Independently acquire any additional skills, concepts, and/or tools necessary for successful project completion.
6. Communicate regularly with the faculty mentor and keep a journal of all work completed in the project.
7. Communicate the results of the project to peers and mentors through an oral presentation and a technical report.

Learning Activities

Week 1: Product Ideas and Resources

The first assignment identifies in an informal report a product or an improvement to a product that should be invented. The student must determine what product will be created, who will use it, what benefit it will provide, and how will success be demonstrated…in other words, the student must perform some market research. The student will be required in the next assignment to demonstrate that there is a need for this product. The student must provide a list of resources which will be used, to help define and describe this new product and the people or companies who will be affected by it directly or indirectly. The student needs to include what the user would request from the system performance. These considerations will help identify system "Requirements" or specifications.

The student must meet with the customer or instructor to discuss the expectations, course requirements as well as being prepared to discuss the ideas more fully. By the end of this week, a proposed product should be chosen by the student.

Week 2: Justification of Need Plus Project Plan and Resource

The student should have discussed their proposed project with the professor by this week. What is now needed is a review of the following:

1. Problem that needs to be solved.
2. People who will be impacted
3. Proposed product description, and
4. Acceptable system or product performance and behavior.

Specifically, the student needs to include in a research of proposed solutions and how it will impact the people and companies. The questions the student needs to answer are:

- Who is affected?
- What are their needs?
- What additional information does this provide about constraints?

After answering those questions, students need to think about what proposed system solution is needed to "pass" the user acceptance test. These additional considerations will help further identify system "requirements" or specifications in the course assignments during the upcoming weeks.

Also, the student needs to provide a description of a problem that is supported by documented information showing that the problem exists as well as considering the level-of-effort in terms of costs, time, money, resources, or human work hours needed to solve a problem. The deliverable also needs to include what the user would expect from a "good" system.

Week 3: Formal Proposal and Project Plan

By this time, the student should have a good understanding of the problem and proposed solution. During the past few weeks, the student should have:

- reviewed and framed the problem statement,
- investigated the anticipated users of the proposed product or service with an identified need,
- performed background research on market demand or needs and
- anticipated resources and constraints to implement the solution.

For this assignment, the student submits a formal proposal, including: the problem statement, product or system description that solved the problem and an initial list of requirements/specifications. The proposal should be submitted at least one day before meeting with the instructor to discuss it.

Week 4: Acceptance Test Plan and Protocol

This test plan and protocol needs to demonstrate that the product does what is required and must relate back to the system requirements and specifications. The acceptance test plan and protocol must include: test setup, the inputs to the system, and the correct response of the system. It should include any information about the actual "workings" or implementation of the system and should not prejudice that actual design by specifying any particular hardware or implementation. At this point, the student should focus on "*what*" the system must do and not on "*how*" to implement the system solution.

Week 5: Block Requirements, Interface, and Review

Students must identify major blocks, or subsystems which will help build the product solution. Students need to answer questions about each of the major areas (or subsystem) of implementation. They must review the guidelines for good requirements and allocate the system functions to the various subsystems.

The student must consider:

- system and subsystem level of requirements/specifications,
- a system and subsystem level tests,
- and interface specifications.

Week 6: Preliminary Design Review (PDR)

The PDR should include all of the information to date, as well as identifying critical issues necessary to finish the project:

1. Definition of the product, requirements/specifications, and the acceptance demonstration that the student intends to provide
2. Fundamental subsystem blocks of the system, and details of signal inputs and outputs between the subsystem blocks
3. Alternate product implementations which were rejected
4. Proposed solutions to individual subsystem block performance requirements.

Week 7: Block Details, Details of Interface Block

The student's activity report should include further details of their interface blocks down to the component level. The student must discuss how the different components of the subsystem relate to each other, and how they work internally. The student needs to update revised acceptance tests and any other revisions based on the PDR.

Week 8: Assess Preparation for Critical Design Review and Design Options

The student must review items completed during the previous weeks before their demonstration in Week 11. In addition, the student must identify and provide further research into the technical choices in the system design and then provide theory and data to support the choice of the best system.

Week 9: Justified Options and Details of Control

The student must support their selected design among the options that were researched. The student should include any details of control. Details of control means to describe the student's proposed functional and system architecture in their system final configuration down to the component level of implementation.

Week 10: Preparation for Critical Design Review and Outline of Oral Presentation to Engineering Faculty

The student meets with instructor to discuss preparation for CDR presentation and demonstration. The discussion must include:

- The product description and must identify users, environment, benefits, and performance
- Specifications that will include the items to be measured to demonstrate the product does what was promised, and the test plan protocol which defines "pass" for each requirement.
- Review of the acceptance test protocol
- Demonstration of the product, in hardware or modeling to show performance and quality of propose product for the final oral presentation
- Input from the "customer" will then be use to provide corrections in the final report detailing:

    1. the product,
    2. specifications,
    3. system diagram including blocks, and interfaces,
    4. interface specifications,
    5. detailed subsystems and
    6. design tradeoffs.

For design tradeoffs, the instructor is looking for what choices did the student make and why. The student must include the rationale of hardware/software selection as well as the design choices that the student eliminated.

Week 11: Final Report and Oral Presentation, Technical Merit

The final technical report is a polished version and compilation of previous work and deliverables that was revised and fine-tuned to produce a professional document. In addition to items for Week 10, the student must include:

- A presentation due on Week 6 defining the product and outline their progress to date (Preliminary Design Review)
- A demonstration of the product or providing sufficient modeling to demonstrate performance and quality of the proposed product (Critical Design Review)
- Other items identified in weekly progress and activity reports
- A final report detailing:

1. the product,
2. specifications,
3. system diagram including blocks, and interfaces,
4. interface specifications,
5. detailed subsystems, and
6. design tradeoffs, that is: what choices did the student make and why including the components, or elimination of design choices and why?

In addition to the above, a systems engineering deliverables, the completed project must consider sufficient scope and technical merit to demonstrate proficiency and expertise in this capstone project. The next section lists several factors and considerations to help stress the importance of technical merit during the execution of the capstone project.

Technical Merit

The engineering department anticipated that a student can satisfactory meet the weekly deliverables from the system engineering process but the project may lack technical depth at the Masters level. To take this into account a percentage weight of 30% for technical merit is considered. Several factors are considered to evaluate and assess the technical depth and merit of a project including:

- Degree of technical difficulty in solving a problem and finding a solution
- Evidence and technical depth of analysis on the solution
- Considerations and technical depth of alternative solutions
- Substantiate proposed solutions with data and facts including technical and economic feasibility
- Evidence and technical depth of synthesis and evaluation tasks from engineering courses in the Master's program
- Evidence and technical depth of newly acquired technical skills not emphasized or taught in the program (also measures degree of independent learning by the student)
- Evidence of integrating information from many sources to gain insight to the problem
- Evidence and technical depth of multiple perspectives for a given problem (economic feasibility, reliability, maintainability, sustainability, safety, etc.).

Example 13. Active Learning: Promoting Curiosity Through a Toaster Teardown

Author: Dr. Michael Swartwout
Course: Freshman Engineering
Learning Activity Approach: Active Learning

Description

When addressing customer needs, engineers need to consider not only the design and manufacturing of the product but also its marketing. This module has students take apart electrical appliances to observe their design, understand how they work and also how marketing choices were made to sell the product to the desired audience group. By "tearing down" a product, one can understand the detailed construction or composition. Mass manufacturing is different than custom manufacturing and the differences that can be seen in the designing and making of the product.

Learning Objectives

1. Investigate technical designs through testing.
2. Analyze how different products are made to understand mass manufacturing.

Learning Activities

The class handout provided for this section was originally developed for toasters; however, it can be modified to investigate different kinds of store bought appliances.

Topic 1: Consumer Motivation

Why buy this appliance? As a team, decide your top-three factors in purchasing the appliance. What are the three things that will make you choose this appliance over another of its kind?

Topic 2: Marketing

Consider the Box. How well does the packaging "sell" the appliance? Specifically, does the packaging address any of the three factors you listed above? Does it address any issues that you had not put in your top three? Based on packaging alone, can you make any guesses about the target audience for this appliance?

Topic 3: Design for Shipping

Open the box and remove the contents. [IMPORTANT: Save everything associated with this appliance. Every plastic wrapper. Every fastener. Every scrap of paper. You will need all of them for a later assignment. Briefly list the items you found in the box, and their function. Why are they there?

Topic 4: Functionality

Write/draw a step-by-step process for turning what is put into the appliance into the output. Identify all the interfaces between the appliance and the outside world. List or draw all the things that the appliance must do to convert the input to the output.

Topic 5: Product Design

Remove the cover, and examine it. What materials are used in making the cover? How is this cover made? Why is it made the way it is? Does it seem overly complicated for its function?

Topic 6: Bread Holder Mechanism Design

Find and expose the mechanism used for holding/ejecting the bread. How does it work? How many parts are needed to operate this mechanism? How is the toast automatically released? Can it be manually released as well as automatically released? (Why would this be a good idea?)

Topic 7: Heater Coil Element Design

Find and expose the heating element(s). How does it work? How does the appliance "know" when to stop? What control does the customer have over the process?

Topic 8: Frame Design

Remove the major parts so that the toaster frame is revealed. How heavy is the frame (relative to the rest of the toaster)? Is it a large or small part of the overall weight? Why is there a frame? Why is the cover/exterior separate from the frame? (In other words, why not just bolt everything to the cover and do away with the frame?)

Topic 9: Maintenance Requirements

What sorts of routine maintenance/cleaning would the customer need to do on this appliance? How does the design of the appliance help or prevent the customer from doing these tasks?

Topic 10: Miscellaneous

What else is in there? Are there other parts that don't seem to participate in the functions listed above? What are their functions?

Example 14. Applying the Lean Start-up Framework: Capstone Design

Authors: Dr. Jose L. Zayas-Castro and Dr. Susana Lai-Yuen
Course: Industrial and Systems Engineering Capstone Design (2 Semesters)
Learning Activity Type: Appling the Learn Start-up Framework

Description

This two-course sequence is a requirement for graduation for all students pursuing a bachelor's degree in Industrial and Management Systems Engineering. This course is team-based and uses the Lean Start-Up framework on which teams of 2–3 students go through the full experience of generating a tech-based idea to address a problem/need. Fundamentally, the teams are required to form a start-up/small company, assess market needs, define a minimum viable product, design the product/solution, formulate the different aspects of manufacturing/production, distribution, costs, and evaluate economical and engineering/technological viability. Therefore, students are expected to integrate knowledge and further their technical, critical thinking, writing, and teamwork skills. The learning activity helps students go through a more realistic experience of identifying a problem to be solved, and discovering the evidence needed to propose and validate an engineering-based solution in a business-like environment. Getting out of the classroom is the focus of the course.

Learning Objectives

1. Investigate the challenges and uncertainties of new idea generation and product development to form a tech-based entrepreneurial small company.
2. Develop collaboration skills necessary for the realization of the start-up/new small company.
3. Develop written and oral communication skills.

Learning Activities

1. *Meyers–Briggs Test*

Students are required to take the Meyers–Briggs test to understand individual personal styles and how it has the potential to affect and influence team performance. This provides self-awareness and is solely for the student's benefit.

2. *Needs Analysis*

Students must determine a technology-based need or a problem to be solved.

3. *Experimentation*

Students must design the experiments (e.g., survey, observation, etc.…) to be tested in an attempt to validate needs and customer assumptions.

4. *Customer Interviews*

Students must "get out of the classroom" to contact and interview customers in an attempt to validate needs and assumptions.

5. *Business Model Canvas*

Students must consider and identify a starting point for the nine components of the business model canvas.

6. *Evaluation*

Students must generate and evaluate the engineering and science related approaches for the proposed solution, addressing opportunities to validate findings.

7. *Prototype*

Students must design and develop a prototype.

8. *Desirability–Feasibility–Viability*

Students must validate and justify the customer desirability, technology feasibility, and business viability of the newly developed design.

9. *Assessment*

There are five major assessment components required throughout each semester to promote practice and feedback.

- Students are required to complete a total of four team-based oral presentations and written reports throughout the semester.
- Individual students are required to submit progress reports evaluating team performance.
- Individual students are required to complete confidential self and peer evaluations at various points.
- Students teams receive interviews by the instructors to assess student learning related to the learning objectives.
- At the end of each semester, student teams develop and display posters providing an overview of start-up/small company.

Example 15. Flipped Classroom "Makecourse": A New Approach to Engineering Discovery

Author: Dr. Rudy Schlaf
Course: Makecourse—Introduction to Engineering Design
Learning Activity Approach: Flipped Classroom "Makecourse"

Description

This course is an elective for all undergraduate students. This course introduces students to practical engineering design skills. Students invent, design, and make a mechatronic object during the semester long course. They learn using professional CAD software, design a microcontroller-based control system, and learn programming code in C++ during the implementation of their project. Each student designs their own project, but they are allowed to collaborate and help each other. This makes each student a team leader as well as a team member. The learning activity helps students to discover their "inner engineers," to build their practical skill set as well as instilling a "can do spirit," which encourages them to make engineering their lifestyle and passion, instead of just "a job."

Learning Objectives

1. Discover and discuss the creative process and the joy of inventing.
2. Identify basic design and innovation methodologies.
3. Create a new invention or design.
4. Apply project and time management skills.

Learning Activities

Summary of Flipped Classroom Approach

The course is taught using the flipped classroom pedagogy (i.e., lectures are outside the classroom). All course materials (how-to videos) are posted on www.makecourse.com and available to the public. All students purchase a standard course kit containing most of the parts needed for their projects and some basic tools (e.g., multimeter, caliper). The instructor and undergrad TAs advise the students during class and office hours and offer one-on-one counseling and project discussions. Students are encouraged to asynchronously watch course videos as needed for their projects. Exams and quizzes are replaced by deliverables. The deliverables are milestones that need to be met on the way to finalizing the student projects. Most deliverables require the authoring of a short video where students report their project progress. The first 10 weeks of the course also feature small Homework projects, encouraging students to explore the provided hardware and build the principal skills required to build their projects.

Weeks 1–4

The first three weeks will be spent in learning basic 3D design skills. We will use Autodesk inventor to design individual parts and assemble them into more complex structures. This segment will be capped by an introduction to dynamic constraints, which allow the simulation of motion processes. Howard Kaplan of the USF Advanced Visualization Center will give a presentation discussing 3D printing. The center has a makerbot farm, which will be used by the students. Students will receive hands-on training with the printers and begin printing first objects.

Weeks 5

In week 5 all students will present their mechanical design component in a short video using their Inventor simulation on their laptop. The designs will be evaluated by the instructors and students will be given feedback if it is necessary to further improve their designs for the project. We will also start exploring the Arduino and basic electronic circuits necessary for the MAKE projects.

Weeks 6–10

Weeks 6–10 will be used to learn the basic skills required to build electronic control systems. All projects will have a mechanical component and an electronic control component involving microcontroller controls, motors, switches, LEDs, etc. This part of the course will focus on getting hands-on experience with programming the Arduino platform, building essential electronic circuits connected to the Arduino outputs and inputs, and the implementation of motorized actuators for driving the mechanical components.

Week 11

In week 11 all students will present the design of their electronic and control setups. The class and the instructors will grade the designs. At this point, the project needs to be fully designed. The remaining weeks of the semester will be used to translate the project into reality and to get ready for the final presentation.

Weeks 12–15

The final weeks of MAKE will focus on guiding the students toward successful completion of their projects. In parallel, we will discuss some more advanced concepts with regard to 3D design and Arduino control systems such as the use of processor interrupts.

Week 16

At the end of the semester all students will present their projects. A video and an Instructable will be submitted in fulfillment of the final deliverable requirement.

Potential Variations

- Makecourse Art: In this course, teams of engineering and art students work together to build functional and artistic focused design projects.
- Makecourse IOT: In this course, students will be introduced to the "Internet of Things" concepts. Students will be encouraged to develop ideas preparing them for the IOT revolution in industry and customer realms.

Example 16. Entrepreneurially Minded Tools: Design Thinking Projects

Organization: Stanford d.School
Website: http://dschool.stanford.edu/
Learning Activity Approach: Entrepreneurially Minded Tools

Description

Stanford's d.School, located in Silicon Valley, is well known for the Design Thinking movement. Here, the design thinking methodology combines analytical and creative approach, drawing on innovation approaches grounded from the engineering and design paradigms. The method encompasses five main iterative and fluid steps including:

- Empathize
- Define
- Ideate
- Prototype
- Test.

The d.School does a wonderful job of explaining the design thinking methodology and provides many free resources for instructors to incorporate the design thinking process into any classroom. Design thinking projects offer a great way to incorporate the entrepreneurial mindset into the middle level non-design engineering courses.

Learning Objectives

1. Investigate and apply creative thinking to course topic related design problems and opportunities.
2. Explain what factors influence design decision-making and who constitute the main design decision makers.

Proposed Learning Activities

The d.School handouts, shown below, offer a great starting point for introducing students to the design thinking process. This exercise can easily be modified to consider a design challenge related to a specific engineering course topic.

1 Interview
8min (2 sessions x 4 minutes each)

Notes from your first interview

Switch roles & repeat Interview 1.

2 Dig deeper
8min (2 sessions x 4 minutes each)

Notes from your second interview

Switch roles & repeat Interview 2.

3 Capture findings 3min

needs: things they are trying to do*
*use verbs

insights: new learnings about your partner's feelings/ worldview to leverage in your design*
*make inferences from what you heard

4 Define problem statement 3min

____
partner name/description

needs a way to ____________________
user's need

Surprisingly // because // but . . .
[circle one]

____

____
insight

5 Sketch at least 5 *radical* ways to meet your user's needs. 4min

write your problem statement above

6 Share your solutions & capture feedback. 8min (2 sessions x 4 minutes each)

Notes

Switch roles & repeat sharing.

d.

7 Reflect & generate a new solution. 3min

Sketch your big idea, note details if necessary!

d.

Example 17. Entrepreneurially Minded Tools: Value Proposition Canvas

Organization: Strategyzer
Website: https://strategyzer.com/canvas/value-proposition-canvas
Learning Activity Approach: Entrepreneurially Minded Tools

Description

The Value Proposition Canvas offers a simple and visual way to understand and communicate the customer needs (what they want and don't want) and design product and service features (identifying how the design will meet customer needs).

Proposed Learning Activities

In any engineering class, students can be assigned or suggest a conceptual innovation or design idea that must be considered with respect to the Value Proposition Canvas components including the product/service bundle and the customer wants and needs.

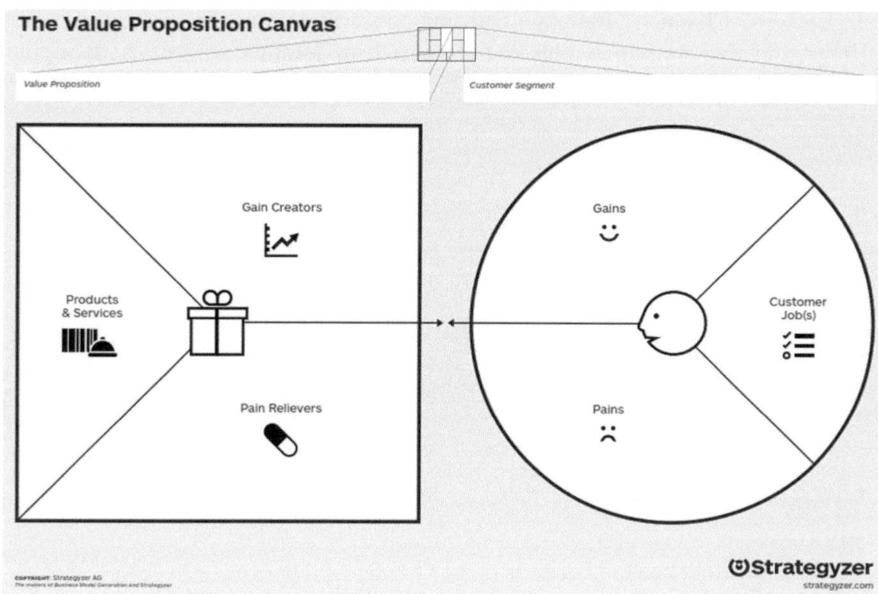

Example 18. Entrepreneurially Minded Tools: Business Model Canvas

Organization: Strategyzer
Website: https://strategyzer.com/canvas/business-model-canvas
Learning Activity Approach: Entrepreneurially Minded Tools

Description

The Business Model Canvas offers a simple and visual way to describe, evaluate, and pivot the business model taking into consideration nine main components: Value Proposition, Customer Segments, Customer Relationships, Channels, Revenue Streams, Cost Structure, Key Resources, Key Partners, and Key Activities.

Proposed Learning Activities

Business Model Canvas

In any engineering class, students can be assigned or suggest a conceptual innovation or design idea that must be considered with respect to the Business Model Canvas components including Value Proposition, Customer Segments, Customer Relationships, Channels, Revenue Streams, Cost Structure, Key Resources, Key Partners, and Key Activities. This is a great starting point for students to then pivot and iterate based on the Test Card and Learning Card shown on the next page.

Business Model Canvas Pivoting (Test and Learn)

The Test Card and Learning Card, shown below, provide an easy way to propose and test assumptions and hypotheses about the business model canvas. These fit well in the "Evaluate" stage of the Discovery-Evaluate-Exploit process.

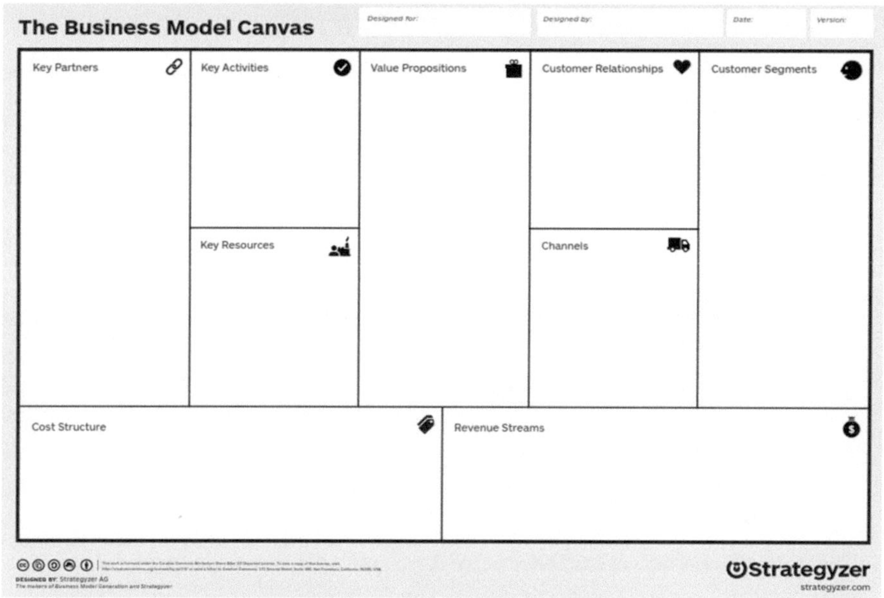
The Business Model Canvas
Designed for:
Designed by:
Date:
Version:
Key Partners
Key Activities
Key Resources
Value Propositions
Customer Relationships
Channels
Customer Segments
Cost Structure
Revenue Streams
DESIGNED BY: Strategyzer AG
Strategyzer
strategyzer.com

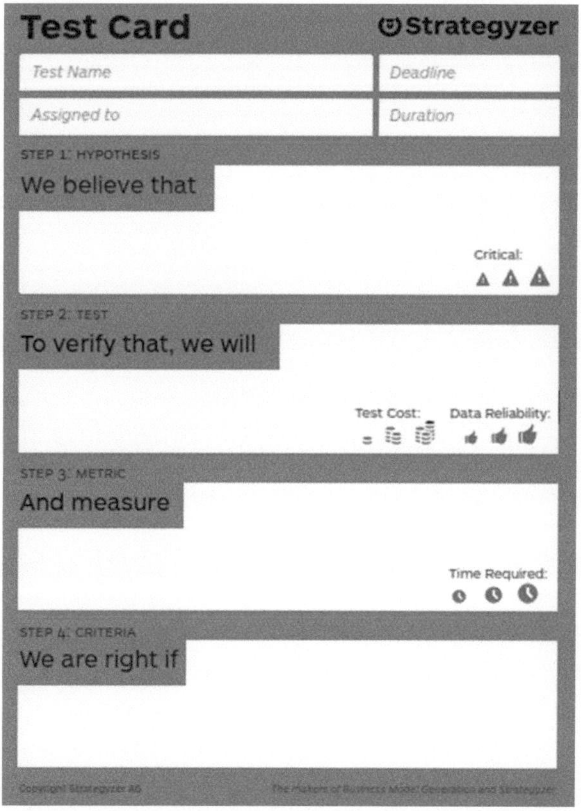
Test Card
Strategyzer
Test Name
Deadline
Assigned to
Duration
STEP 1: HYPOTHESIS
We believe that
Critical:
STEP 2: TEST
To verify that, we will
Test Cost:
Data Reliability:
STEP 3: METRIC
And measure
Time Required:
STEP 4: CRITERIA
We are right if
Copyright Strategyzer AG
The Makers of Business Model Generation and Strategyzer

Learning Card
Strategyzer
Insight Name
Date of Learning
Person Responsible
STEP 1: HYPOTHESIS
We believed that
STEP 2: OBSERVATION
We observed
Data Reliability:
STEP 3: LEARNINGS AND INSIGHTS
From that we learned that
Action Required:
STEP 4: DECISIONS AND ACTIONS
Therefore, we will
Copyright Strategyzer AG
The makers of Business Model Generation and Strategyzer

Example 19. Entrepreneurially Minded Tools: How to Build a Start-up with Steve Blank

Organization: Udacity
Website: https://www.udacity.com/course/how-to-build-a-startup–ep245
Learning Activity Approach: Entrepreneurially Minded Tools

Description

This course is offered for free through Udacity, a massive open online course, and was developed and produced by Steve Blank, a serial-entrepreneur and professor recognized for his work in the Lean Start-up movement. Students will learn the business viability skills necessary to bring an idea from a conceptual design to market. At the end of the course, students will be able to do the following:

1. Describe customer wants and needs through active listening and customer engagement.
2. Evaluate customer feedback to make a product or service stronger through marketing and business model modifications.
3. Explain the customer relationship management lifecyle including (a) Get, (b) Keep, and (c) Grow.
4. Identify key resources, key partners, key activities, and the distribution channel required to produce and deliver the product or service to the customer.
5. Calculate direct and indirect expenses required to produce and deliver the product or service to the customer.

This course is comprised of a series of short videos and activities based on the following topics:

Lesson 1: What We Now Know

- History of the Corporation
- Start-ups Are Not Smaller Versions of Large Companies
- Waterfall Development
- Customer versus Product Development
- Entrepreneurial Education.

Lessons 1.5A and 1.5B: Business Models and Customer Development

- Value Proposition
- Customer Segments
- Revenue Streams
- Key Resources
- Customer Development Processes
- Minimum Viable Product
- Market Opportunity Analysis.

Lesson 2: Value Proposition

- Value Proposition and the Minimum Viable Product
- Customer Archetype
- MVP Physical && Web/Mobile
- Common Mistakes With Value Proposition

Lesson 3: Customer Segments

- Product Market Fit
- Rank and Day in the Life
- Multiple Customer Segments
- Market Types Introduction: Existing, Resegmented, New, Clone
- Consequences of Not Understanding a Market

Lesson 4: Channels

- Distribution Channels Overview
- Web Distribution
- Physical Distribution
- Direct Channel Fit
- Indirect Channel Economics
- OEM Channel Economics

Lesson 5: Customer Relationships

- Paid Demand Creation
- Earned Demand Creation
- Get Physical
- Viral Loop
- Web Customer Acquisition Costs

Lesson 6: Revenue Models

- How Do You Make Money
- Revenue Streams and Price
- Direct and Ancillary Models
- Common Start-up Mistakes
- Market Types and Pricing
- Single and Multiple Side Markets
- Revenue First Companies
- Market Size and Share

Lesson 7: Partners

- Partner Definition
- Partner Resources
- Partner Types
- Greatest Strategic Alliance
- Joint Business Development

Lesson 8: Resources, Activities, and Costs

- Four Critical Resources
- Financial Resources
- Human Resources
- Qualified Employees and Culture
- Intellectual Property Overview.

Proposed Learning Activities

Example 1: Online Discussions

Entrepreneurially minded online discussions provide a great learning opportunity to promote curiosity and connection of course topics to real-world entrepreneurial-minded applications. Here are some examples incorporating the Udacity videos into an online discussion:

- Watch Lesson 2: Value Proposition. Select an innovation related to the course topic. Identify which components make up the minimally viable product. Analyze what that innovation might reduce, raise, eliminate, or create. Then, write a value proposition statement based on the analysis.
- Watch Lesson 2: Value Proposition. Select an innovation connected to your course content and list all of the potential channels for marketing and distributing it. Address the following statements: (a) Many people think that social media is the best way to market new products. Is this true for the innovation you were assigned or selected? Why or why not? Describe the potential benefits and challenges of using social media to market your innovation. (b) Many people think major online distributors such as Amazon or eBay are the best way to sell new products. Is this true for the innovation you were assigned or selected? Why or why not? Describe the potential benefits and challenges of using online retail to sell your innovation.
- Watch Lesson 6: Revenue Models. Select an innovation related to the course content. List all the possible revenue streams. Address the following statement: (a) Many entrepreneurs find revenue streams other than outright sale of their product attractive (such as, rent/lease, subscription, licensing). Describe the potential benefits and challenges of using these alternative revenue streams for their selected/assigned innovation.
- Watch Lessons 1.5A and 1.5B: Business Models and Customer Development. Select and innovation related to the course topic. Search online for similar products an develop a list of competitors. Describe the landscape of competition for the innovation. Are there a lot of competitors providing similar products or services? Are there any large competitors likely dominating the market share? How entrenched are the competitors? Are the competitors start-ups or long-standing companies? Write a short description explaining how the selected innovation provides something different than its competitors.

Example 2: Value Creation Projects

Value creation projects connect the traditional scientific method and engineering design process to business and marketing through a focus on goals rather than problems. This iterative process encourages engineering students to think outside the box, beyond the current state of parameters to consider the potential that the future holds.

The Udacity videos can offer a great compliment to Value Creation Projects. First, Lesson 1: What We Now Know, can provide a background and motivation for how and why these types of projects matter in the real world. Second, Lessons 1.5A and 1.5B: Business Models and Customer Development, can provide a general overview of the value creation process. Lastly, Lesson 2: Value Propsition, can provide a more detailed perspective of what exactly constitutes value creation in a real-world context.

Furthermore, Value Creation Projects that incorporate Design Thinking, the Value Proposition Canvas, and/or the Business Model Canvas, can benefit from students watching all or parts of the Udacity course.

MIX
Papier aus verantwortungsvollen Quellen
Paper from responsible sources
FSC® C105338

If you have any concerns about our products,
you can contact us on
ProductSafety@springernature.com

In case Publisher is established outside the EU,
the EU authorized representative is:
Springer Nature Customer Service Center GmbH
Europaplatz 3, 69115 Heidelberg, Germany

Printed by Libri Plureos GmbH
in Hamburg, Germany